DECISION MAKING FOR SUSTAINING OPERATIONAL EXCELLENCE IN NUCLEAR POWER PLANT OPERATING ORGANIZATIONS

The following States are Members of the International Atomic Energy Agency:

AFGHANISTAN	GEORGIA	PAKISTAN
ALBANIA	GERMANY	PALAU
ALGERIA	GHANA	PANAMA
ANGOLA	GREECE	PAPUA NEW GUINEA
ANTIGUA AND BARBUDA	GRENADA	PARAGUAY
ARGENTINA	GUATEMALA	PERU
ARMENIA	GUINEA	PHILIPPINES
AUSTRALIA	GUYANA	POLAND
AUSTRIA	HAITI	PORTUGAL
AZERBAIJAN	HOLY SEE	QATAR
BAHAMAS, THE	HONDURAS	REPUBLIC OF MOLDOVA
BAHRAIN	HUNGARY	ROMANIA
BANGLADESH	ICELAND	RUSSIAN FEDERATION
BARBADOS	INDIA	RWANDA
BELARUS	INDONESIA	SAINT KITTS AND NEVIS
BELGIUM	IRAN, ISLAMIC REPUBLIC OF	SAINT LUCIA
BELIZE	IRAQ	SAINT VINCENT AND
BENIN	IRELAND	THE GRENADINES
BOLIVIA, PLURINATIONAL	ISRAEL	SAMOA
STATE OF	ITALY	SAN MARINO
BOSNIA AND HERZEGOVINA	JAMAICA	SAUDI ARABIA
BOTSWANA	JAPAN	SENEGAL
BRAZIL	JORDAN	SERBIA
BRUNEI DARUSSALAM	KAZAKHSTAN	SEYCHELLES
BULGARIA	KENYA	SIERRA LEONE
BURKINA FASO	KOREA, REPUBLIC OF	SINGAPORE
BURUNDI	KUWAIT	SLOVAKIA
CABO VERDE	KYRGYZSTAN	SLOVENIA
CAMBODIA	LAO PEOPLE'S DEMOCRATIC	SOMALIA
CAMEROON	REPUBLIC	SOUTH AFRICA
CANADA	LATVIA	SPAIN
CENTRAL AFRICAN	LEBANON	SRI LANKA
REPUBLIC	LESOTHO	SUDAN
CHAD	LIBERIA	SWEDEN
CHILE	LIBYA	SWITZERLAND
CHINA	LIECHTENSTEIN	SYRIAN ARAB REPUBLIC
COLOMBIA	LITHUANIA	TAJIKISTAN
COMOROS	LUXEMBOURG	THAILAND
CONGO	MADAGASCAR	TOGO
COOK ISLANDS	MALAWI	TONGA
COSTA RICA	MALAYSIA	TRINIDAD AND TOBAGO
CÔTE D'IVOIRE	MALI	TUNISIA
CROATIA	MALTA	TÜRKİYE
CUBA	MARSHALL ISLANDS	TURKMENISTAN
CYPRUS	MAURITANIA	UGANDA
CZECH REPUBLIC	MAURITIUS	UKRAINE
DEMOCRATIC REPUBLIC	MEXICO	UNITED ARAB EMIRATES
OF THE CONGO	MONACO	UNITED KINGDOM OF
DENMARK	MONGOLIA	GREAT BRITAIN AND
DJIBOUTI	MONTENEGRO	NORTHERN IRELAND
DOMINICA	MOROCCO	UNITED REPUBLIC OF TANZANIA
DOMINICAN REPUBLIC	MOZAMBIQUE	UNITED STATES OF AMERICA
ECUADOR	MYANMAR	URUGUAY
EGYPT	NAMIBIA	UZBEKISTAN
EL SALVADOR	NEPAL	VANUATU
ERITREA	NETHERLANDS,	VENEZUELA, BOLIVARIAN
ESTONIA	KINGDOM OF THE	REPUBLIC OF
ESWATINI	NEW ZEALAND	VIET NAM
ETHIOPIA	NICARAGUA	YEMEN
FIJI	NIGER	ZAMBIA
FINLAND	NIGERIA	ZIMBABWE
FRANCE	NORTH MACEDONIA	
GABON	NORWAY	
GAMBIA, THE	OMAN	

The Agency's Statute was approved on 23 October 1956 by the Conference on the Statute of the IAEA held at United Nations Headquarters, New York; it entered into force on 29 July 1957. The Headquarters of the Agency are situated in Vienna. Its principal objective is "to accelerate and enlarge the contribution of atomic energy to peace, health and prosperity throughout the world".

IAEA-TECDOC-2108

DECISION MAKING FOR SUSTAINING OPERATIONAL EXCELLENCE IN NUCLEAR POWER PLANT OPERATING ORGANIZATIONS

INTERNATIONAL ATOMIC ENERGY AGENCY
VIENNA, 2025

COPYRIGHT NOTICE

For further information on this publication, please contact:

Nuclear Power Engineering Section
International Atomic Energy Agency
Vienna International Centre
PO Box 100
1400 Vienna, Austria
Email: Official.Mail@iaea.org

© IAEA, 2025
Printed by the IAEA in Austria
December 2025
https://doi.org/10.61092/iaea.lve6-afwm

IAEA Library Cataloguing in Publication Data

Names: International Atomic Energy Agency.
Title: Decision making for sustaining operational excellence in nuclear power plant operating
    organizations / International Atomic Energy Agency.
Description: Vienna : International Atomic Energy Agency, 2025. | Series: IAEA-TECDOC
    ISSN 1011-4289 ; no. 2108 | Includes bibliographical references.
Identifiers: IAEAL 25-01801 | ISBN 978-92-0-127025-2 (paperback : alk. paper) |
    ISBN 978-92-0-126925-6 (pdf)
Subjects: LCSH: Nuclear power plants — Decision making. | Nuclear power plants —
    Management. | Nuclear power plants — Safety measures.

FOREWORD

Operating and managing a nuclear power plant involves making decisions, both large and small, at every level, from planning and design to training and resource management. These decisions can directly affect plant safety and performance, as well as the motivation of the workforce. The potential benefits and risks associated with decisions in the nuclear power generation industry can be of far greater consequence than in other business settings, and past nuclear events and accidents provide opportunities to learn how decisions can be improved.

To ensure the best outcomes, nuclear power plant operators need to gather information continually and systematically, analyse it, make sound decisions based on the data, develop appropriate initiatives, and implement them successfully. This publication is intended to equip senior managers with practical processes, mechanisms and tools for making timely and effective decisions.

Although nuclear power plant operators face many types of decisions in the context of their work, this publication focuses on the strategic and tactical decisions that have the most significant long term effect on plant safety and performance. Strategic and tactical decision making within these organizations is often complex, influenced by internal and external factors such as national and organizational culture, social and economic conditions, government policies, and the positions of regulators and stakeholders. The interplay between these factors, and their combined effect, can lead to pitfalls for decision makers, some of which are fairly obvious and others less readily apparent.

One common example is heuristics: mental shortcuts or 'rules of thumb' that reduce the cognitive load on decision makers. Decisions based on heuristics are often quick, but they also tend to be overly optimistic and relatively inflexible, rather than well considered, critically examined and flexibly adaptable as circumstances evolve.

Senior managers need to be aware that such pitfalls are an inherent risk in complex decision making. This publication presents concrete examples of decision making challenges encountered within nuclear power plant operating organizations and provides tools, initiatives and strategies to overcome them, along with best practices and lessons learned from IAEA Member States.

The IAEA officer responsible for this publication was A. Kawano of the Division of Nuclear Power.

CONTENT

1. INTRODUCTION

1.1. BACKGROUND

Decisions related to nuclear power plants (NPPs) can significantly impact the community both today and for many decades to come. While the scale and potential impact of relevant enterprises are substantial, the lengthy plant life cycle—often exceeding 100 years from construction to decommissioning—underscores the importance of having effective decision making (DM) processes in place from the outset of any new build project and maintaining them throughout the plant's life cycle. It is not only the decisions related to plant design that are crucial early in the life cycle but also those concerning people, processes, culture, and leadership. These factors determine the success of the nuclear power station and, once established, are often more challenging to change.

IAEA Nuclear Energy Series No. NR-G-3.1 [1] on Sustaining Operational Excellence at Nuclear Power Plants supports senior executives of NPP owner/operator organizations by providing strategic responses to current business challenges and effective measures to sustain high-performance levels. This publication raises four vital core principles for sustaining performance. Those are:

- Integrated plant knowledge;
- Culture for learning and performance improvement;
- Pride in one's profession and work;
- Reliability of the equipment.

Soundness in leadership, organizational culture, and decisions made based on them would enable and ensure these four. For example, integrated plant knowledge cannot be developed and maintained without management decisions and commitment to knowledge management. This commitment includes keeping design information accurate and ensuring there is a well-structured continuous training system on plant design, the licensing basis, operational experiences and international standards. Additionally, equipment will not be kept reliable without a well-thought-out maintenance strategy.

IAEA publications have yet failed to include[1] comprehensive guidance to describe the basics of DM, such as a structured DM process, possible pitfalls, methods and tools to support DM. In order to help the owner/operator of NPPs install a DM system within their management system, which everyone understands and adheres to, this publication discusses the necessary considerations for effective DM with specific examples based on an understanding of current challenges and opportunities and learnings of past events and experiences.

In general, the decisions in the nuclear energy industry include those related to design, construction, operation, decommissioning of NPPs and waste disposal. However, this document focuses on support of operational excellence, although the basics of DM, as mentioned above, more or less apply.

---

[1] Section 3.4 of TECDOC-2038 [2] on Institutional Strength in Depth (ISiD) is dedicated for DM as one of the key attributes of ISiD, and discusses some key elements in DM (system, diverse view, risk informed and organizational traits). However, because of the nature of this document that deals with "institutional strength", it does not touch upon such matters as process, pitfalls, methods/tools, good practices etc.

1.2. OBJECTIVE

This publication aims to provide senior managers of NPP owner/ operating organizations with specific and practical suggestions and examples for their effective DM to enhance safety further, improve performance, maintain economic competitiveness and sustain operational excellence. Possible resolutions of processes, tools, and frameworks are given.

1.3. SCOPE

DM related to nuclear power falls into three categories:

- Strategic (such as long-term, cooperate policy, and technology selection for new nuclear build);
- Tactical (such as programmed financing, plan to hire new talent and train);
- Operational (on an hourly, daily or weekly basis, dealing with short-term plant operational details, daily resource allocation, inventory control, etc.).

While strategic DM provides the organization with its direction to go forward, and tactical DM how the direction can be implemented and achieved, operational DM is more short-term and directly related to daily plant operation and management. Considering recent changes in the nuclear business environment and reflecting the lessons learned from experience, the discussions in this publication focus on strategic and tactical DM and go beyond plant operation and management, considering various external factors (economic, political, societal and environmental)[2].

DM under serious emergency (beyond abnormal) is not in the scope of this publication, and this is discussed in separate documents like Refs. [4, 5] by the IAEA Incident and Emergency Centre (IEC).

The publication describes processes, mechanisms, tools, and techniques for ensuring that strategic and tactical decisions are made and implemented effectively, efficiently, and on time at all levels of the organization.

The practices described in this publication may also be applied to other nuclear stakeholders such as manufacturers, vendors, contractors, government organizations, regulatory bodies, technical support organizations, and national and international industry organizations.

1.4. STRUCTURE

This publication is comprised of eight sections.

---

[2] To assist Utilities in threshold conditions, WANO has also developed "Principles for Effective Operational Decision-Making" [3], which has three scenarios: (1) shift manager and control room team make immediate responses to abnormal conditions, (2) station managers make decisions in response to degraded conditions that often fall below action thresholds defined in license documents or that are unclear in existing procedures, (3) utility executives make decisions that have the potential to impact the enterprise. This publication covers primarily the scenario other than scenario (1), that is discussed in other documents and the relevant DM process is already proceduralized or, if quick action is required, automated.

Section 1 sets out the need for the publication, i.e., why the publication has been produced, the objectives and the purpose of the publication, the bounds of the publication, the users and how it has been structured to allow ease of use.

Section 2 discusses the challenges faced by owner/operators in making decisions.

Section 3 describes the basics of DM, including a standard DM process, how the human element can prevent making correct decisions, DM style, and organizational structures for DM.

Section 4 discusses risk assessment and risk informed decision making (RIDM) from several international perspectives.

Section 5 describes how organizations make decisions, including the important role of boards and senior managers, DM structures, knowledge management, leadership and culture.

Section 6 describes the multiple domains where decisions are made and key considerations in those domains that senior managers need to address in their strategic planning and on the issues related to supply chain, government policy, economics, and social and stakeholder affairs.

Section 7 provides examples of tools and methods that can assist in making decisions.

Section 8 summarizes and concludes the publication.

The structure and flow of the publication are also shown in Figure 1.

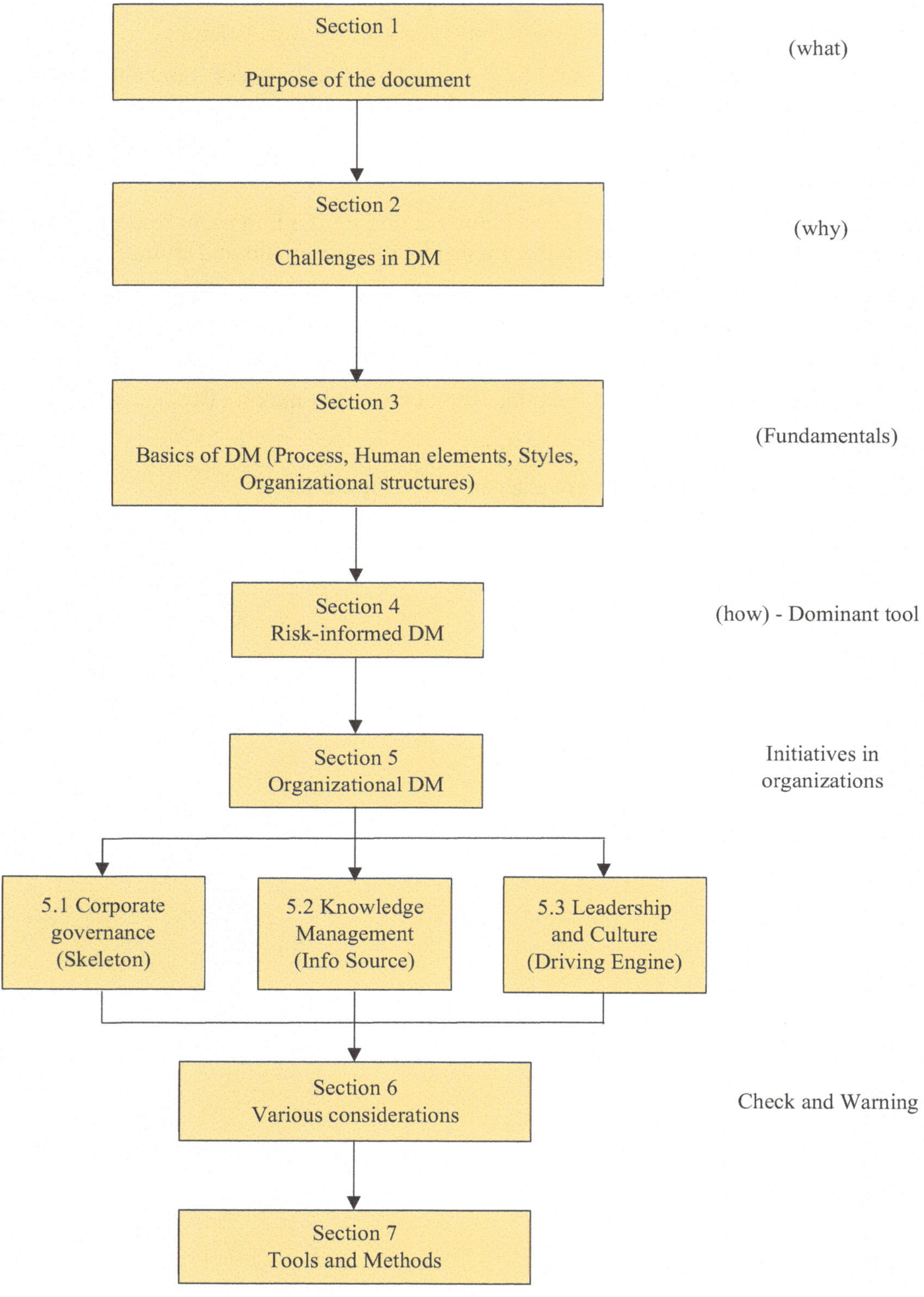

FIG. 1.Structure and flow of the publication

2.   OWNER/OPERATOR CHALLENGES IN DECISION MAKING

The nuclear business requires senior managers and other decision makers to make challenging or tough decisions regularly. Several issues and challenges have arisen due to inappropriate DM. This section discusses the challenges that prevent decision makers from making a right choice, in other words, why they err in DM. The challenges include:

— Insufficient information;
— DM processes and governance;
— Uncertainties;
— Complexity;
— Time pressure;
— Business conditions;
— Organizational and national culture;
— Human factors.

2.1.   INSUFFICIENT INFORMATION

Wise decisions need to be supported by a hierarchy of sufficient and correct data, information, and knowledge [6]. Lack of knowledge for DM can arise from the following:

— Unavailability of correct information on the NPP (design, behaviour, risks, licensing basis, operational experience) surrounding the issue and unreadiness of its use by decision makers;
— Poor options to choose from due to lack of knowledge of plant design, latest findings from research and study and unclear understanding of consequences of each option in terms of effectiveness in achieving the goal, adverse effects, or cost-benefits;
— Lack of collection and consideration of diverse views, including those from different disciplines;
— Lack of information on best practices for tackling the same or similar issue (through participation in Owners Groups, Communities of Practice, standards groups, or similar organizations).

Deferral of DM under a lack of information may not help the situation if the prospect of new information emerging is low or if the need to make the decision is urgent. Questioning attitude can be critical to verify that complete and correct information is being utilised in the DM process.

2.2.   DM PROCESS AND GOVERNANCE

A lack of structured processes (who decides what and how) to make routine and key decisions can challenge some organizations. Decisions may not be made at the right level or be made using heuristics without adequate information or clearly defined decision criteria to make a correct decision. Peer pressure, groupthink, or not valuing the views of less extroverted persons may lead to poor decisions.

Ineffective oversight methods, lack of clarity in authority levels, poorly written procedures, and poor training in DM practices can contribute to poor decisions.

Note: For nuclear, weakness within the design-authority function can contribute to inadvertent poor decisions related to the design basis (refer to INSAG-19 [7]).

2.3. UNCERTAINTIES

Uncertainties can be described as aleatoric (due to inherent randomness that cannot be controlled) or epistemic (due to lack of knowledge or incomplete understanding). Aleatoric uncertainty cannot be reduced, but epistemic uncertainty can often be reduced by better data collection, improved models, or increased knowledge.

Some key uncertainties relevant to nuclear facilities include the impacts of natural hazards, regulatory changes, future innovations, societal norms, stakeholder opinions, government policy, and demand changes. Recurrence periods may be estimated through observation or historical data, but significant uncertainty remains. Due to the long life of NPPs, operating organizations need to consider the potential long-term impacts of climate change [8] (higher rainfall, flood levels, temperatures, weather events).

How to treat uncertainties in PSA for RIDM is discussed extensively in Ref. [9].

2.4. COMPLEXITY

Some decisions are extremely complex and may be impacted by many factors, such as uncertainties, stakeholder pressures, supplier constraints, economic concerns or other issues. Other challenges (e.g., time pressures or lack of information) may contribute to poor decisions for complex problems by disturbing deliberate systemic thinking.

2.5. TIME PRESSURE

Time pressures, often related to economic or production constraints, decrease the decision time available to prepare for and make correct decisions. Individuals tend to rely more on heuristic or biased thinking and incomplete knowledge and may make more risky or incorrect decisions unless other measures are in place.

2.6. BUSINESS CONDITIONS

An organization's business, economic and social situation influences a decision maker's thinking. Such influences may manifest as time pressures, increased attention to cost considerations (leading perhaps to a lack of investment in people, maintenance, or equipment), and inattention to environmental, social or other concerns. Other considerations can include resource availability, government mandates [10], regulations, fiscal policy, incentives, exchange rates, corruption, or bureaucracy, which can also impact DM.

2.7. ORGANIZATIONAL AND NATIONAL CULTURE

Organizational and national culture can more or less impact DM in many ways, including:

- Who is involved in the DM process (top-down vs collaborative cultures, degree of deference to authority, competence and knowledge of decision makers);

- How fast decisions get made (cultures that value agility and speed versus deliberate and cautious DM);
- Criteria used for DM (e.g., efficiency, innovation, risk taking versus risk aversion, the value of knowledge and evidence-based decisions, mission/vision/values of the organization, community involvement/acceptance);
- Risk tolerance (e.g., encouragement of innovation vs. being risk averse);
- Ethics (shared ethical values and norms encourage decisions aligned with those values and norms);
- Sectionalism/siloing (the degree to which parts of an organization inform and communicate with each other; degree of diversity in the DM team);
- Performance Improvement culture (the degree to which learning and corrective actions are valued, degree of independent oversight in place);
- Compliance culture (the extent to which personnel follow defined procedures and processes for effective DM).

Opportunities for correct decisions increase as decisions are made by recognizing these cultures/traits.

2.8. HUMAN FACTORS

Constraints that impact human DM, such as limited time, imperfect information or knowledge, restricted information processing capacity, or lack of team communication, affect DM. These aspects are more fully described later in Section 3.3.

3. DECISION MAKING BASICS

This section describes the basics of DM, including a standard rational DM process, how the human element impacts DM, DM styles and organizational structures related to DM. As Olivier Sibony says [15], "A Good Decision Is a Decision Made the Right Way."

3.1. STRUCTURED DECISION MAKING PROCESS

People tend to make decisions aligned with their personal preferences. Decisions are expected to be fully logical and consistent with rational behaviour. Fig. 2. shows the typical steps of such a structured DM process.

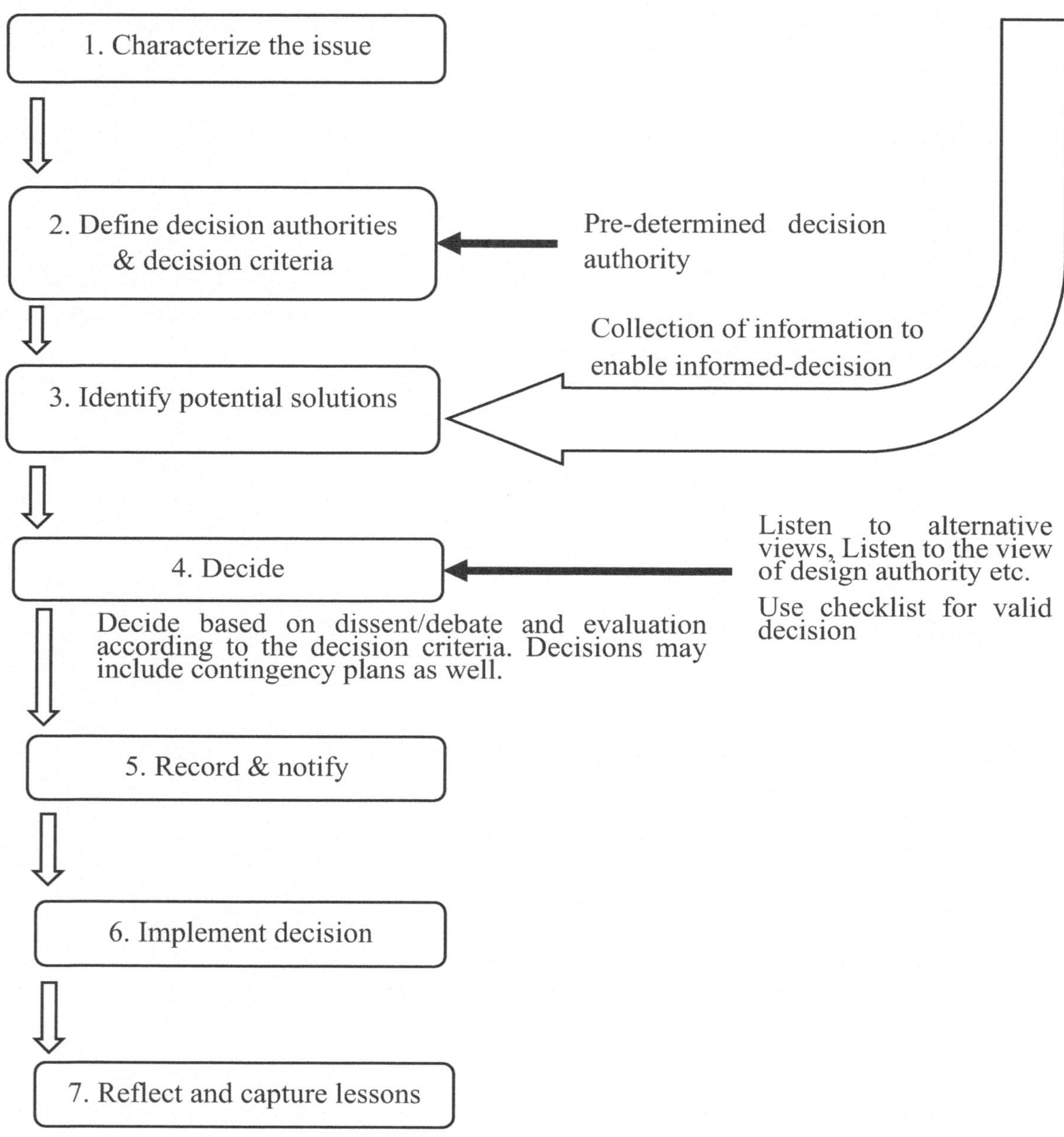

FIG. 2. Decision Making Steps

The crucial aspects of this process are:

- Characterizing the issue: Clearly defining the decision that needs to be made and define a clear timeline (See Section 7.5).
- Collecting information: Ensuring decisions are based on sufficient information, avoiding intuition-based "betting."
- Defining decision authorities: Deciding in advance who will make typical decisions and their levels of authority (accountability, responsibility, ownership; see Section 5.1.2). Factors influencing decision makers include organizational hierarchy, decision impact, and time pressures.
- Defining decision criteria: Establishing decision criteria, weightings and accountabilities. See Section 7.1.1 for some methods to determine criteria and weightings.
- Identifying potential solutions: Develop a set of alternatives for the decision, considering risk and contingency plans.
- Deciding: Selecting the preferred alternative according to the defined decision criteria. Alternate views need to be considered; guidance checklists can assist with this (see Appendix V).
- Recording and notifying: Documenting the decision and contingency plans and informing affected stakeholders.
- Implementing: Taking steps to implement the preferred alternative and monitor its progress and outcomes (see section 3.2 for further details on this step).
- Reflect & capture lessons: Reflecting on good and bad decision experiences and making necessary changes to process, criteria and checklists (see Section 3.2 and Appendix V). Going back to the upstream steps in a flexible manner, in case it is necessary and possible to change decisions.

Section VII.2.1 in Appendix VII describes a similar DM model adopted by EDF in France, Section VII.3 in Appendix VII describes a structured DM model developed in Germany, FOR-DEC, and Section VII.6.1 in Appendix VII describes a structured DM model adopted in the Tennessee Valley Authority (TVA) of the USA, Decision Gates for Advanced Nuclear.

3.2. DECISION IMPLEMENTATION AND FOLLOW-UP

Good decisions can be harmed by poor implementation. Once the authorities for making decisions have been defined and the decision made, the decision needs documentation, stakeholder notification, implementation and follow-up. Organizations need to implement mechanisms for these efforts, using practices from good project, programme and change management principles. The behaviours of the management team involved or impacted by the decision are critical to implementation. Some important aspects of decision follow-up include [11]:

- Support for the decision once made, regardless of personal viewpoint.
- Proactively communicating the process and outcome of the decision, especially to those most affected by the decision.
- Planning for execution and allocation of appropriate resources.
- Holding oneself and each other accountable for execution.

– Tracking the outcome to learn and adjust, paying particular attention to the impact on disparities and advancing equity.
– Not reopening decisions unless significant factors have changed.

Oversight measures such as performance measurement systems, feedback loops, audits and post-implementation reviews can assist with ensuring decisions are correctly implemented and identifying lessons learned and best practices to enhance future decision implementation approaches. Encouraging organizational learning from past decisions helps improve future decision quality. Section 5.1.4 of this publication discusses the role of management systems in establishing DM processes and oversight methods.

Successful decision implementation often requires skilled change management and the application of project management skills to keep track of the necessary activities. IAEA Nuclear Energy Series publication No. NG-T-1.6 [12] provides guidance on project management of nuclear power plant projects. TECDOC-1226 [13] discusses managing change in nuclear utilities.

3.3. HUMAN ELEMENTS IN DECISION MAKING - HEURISTICS, COGNITIVE BIASES AND TRAPS

Humans face limitations in their DM. Even the smartest and most experienced individuals make mistakes. Constraints such as limited time, imperfect information or knowledge, restricted information processing capacity, or lack of team communication affect DM.

When time is scarce, uncertainty is high, or with insufficient information, decision makers rely on mental shortcuts, leading to biased judgements [14]. Heuristics – intuitive ways of obtaining a quick answer through experiences, preconceptions or "rules of thumb"- plays a role. Heuristic-driven cognitive biases can lead to excluding valuable information, misinterpreting or ignoring risks, and avoiding feasible solutions or other negative impacts. They can create traps and lead to mistakes in decisions. Table 16 describes the DM traps proposed by Olivier Sibony [15] with specific examples from the nuclear industry. Appendix I describes the heuristics and biases that create these traps. Appendix II discusses the analysis of DM related to the Fukushima Daiichi Accident, including the aspects of heuristics, biases and traps.

TABLE 1. DECISION TRAPS (ADAPTED FROM REF. [15])

Trap Name	Description	Example
Storytelling trap	Decision makers are swayed by a compelling story without examining the evidence, often aligning with preconceived notions or from a respected or charismatic individual.	Unbalanced, fear-based narratives may influence decisions on nuclear power.
Imitation trap	Decision makers may try to emulate the 'best practices" of a successful company without analysing its applicability.	NPP's senior management emulate the tactics of the legendary executive without considering differences in the surrounding environment.
Intuition trap	Intuition is based on the rapid recognition of a situation that has already been experienced and memorized. Decision makers need to be wary of self-assured experts who place too much faith in their intuition without relevant experiences. Especially, intuition is a poor guide to strategic decision making, since such opportunities may be rare.	A very experienced foreperson in the maintenance division at NPP may trust their good feeling and, even under an unexperienced new situation, tends to make a quick judgement from intuition, especially when time is pressing.

TABLE 1. DECISION TRAPS (ADAPTED FROM REF. [15])

Trap Name	Description	Example
Overconfidence and fixed idea traps	In general, people overestimate their capability and experience of success. However, due to overconfidence in their capabilities and excessive faith in their own predictions, decision makers tend to overlook the fact that success requires many favourable circumstances.	Due to a high level of component reliability at the NPP with its reputation, management may think accidents will not happen there because component reliability ensures safety.
Inertia trap	Decision makers may prefer not to change previous decisions or approaches and remain with the status quo by convincing stakeholders that past investments or efforts were not in vain.	Many projects with high sunk costs have continued to be spent even when not viable.
Risk perception trap	Decision makers avoid exploring new concepts or risks due to excessive aversion to loss or the potential for losses or bad outcomes that are deemed more dominant than the potential for gains or good outcomes. High uncertainty in the probability of the relevant threat happening may also contribute to it and let them take a qualified risk.	Stopping NPP operations to address safety concerns through large-scale modifications causes a significant loss of revenue, which can intimidate decision makers. Also, the high uncertainty in the probability of the threat's happening may lead to no or delayed decision.
Time horizon trap	Decision makers focus on short-term results (e.g., stock price, annual incentives, etc.) instead of long-term benefits.	An NPP may defer preventive maintenance to save current expenses to the detriment of long-term reliability.
Groupthink trap	Decision groups may accept a viewpoint or conclusion representing a perceived group consensus and suppressing hidden concerns. A region or a country of highly homogeneous society or an organization with long shared experience may have a high tendency to fall into this trap.	The common new methodology for evaluating tsunami height proposed by the Japanese Society of Civil Engineers was well accepted and implemented among all the electric utilities in Japan, including TEPCO, regardless of the fact that the potential existence of tsunami sources along the Japan trench off the shore of Fukushima was not deliberately counted nor discussed before the Fukushima Daiichi Accident.
Conflict of Interest trap	Judgement is biased if someone's self-interest (money, promotion, reputation) influences decision makers.	A certain division of an NPP organization tries to justify increased funding to it against the whole benefit of the NPP.

3.4. DECISION MAKING STYLE

Organizations can adopt many DM styles [16, 17]. Table 2 describes the styles, advantages, and disadvantages of methods to make individual decisions. Organizations often combine elements of these styles based on the situation and their culture. A top-down approach is usually used for an organization that is in recovery or has experienced poor performance. In contrast, the same approach would impair the ability of a high-performing organization. The key is finding the right balance between speed, buy-in, proximity to work, and quality of decisions.

TABLE 2. INDIVIDUAL DECISION MAKING STRUCTURES AND PROCESSES

Decision Making Style	Description	Advantages	Disadvantages
Top-Down	Decisions are made at the top management level and cascaded down the organization.	Decisions are made quickly by skilled leaders/managers	Low proximity to the work being done, potentially leading to suboptimal decisions. Low buy-in and empowerment from employees carrying out the decisions.
Consensus	Decisions are made through a collaborative process involving input and agreement from all stakeholders.	High buy-in as everyone's voice is heard	Slow process as consensus takes time. Can lead to groupthink or decisions based on politics rather than merit.
Distributed	The person closest to the work decides after seeking input from relevant stakeholders.	Decisions made by those most familiar with the context. Empowers individuals and drives engagement.	Requires strong DM skills at all levels. Can be slower if decision makers lack experience.
Rational	A systematic approach involves defining the problem, gathering information, generating alternatives, evaluating them, selecting the best option, implementing it, and reviewing the outcome. Steps include: – Identify the decision to be made. – Gather relevant information. – Generate possible solutions. – Evaluate the alternatives. – Select the best option. – Implement the decision. – Review and learn from the outcome.	Comprehensive process that reviews all information.	Can be time-consuming for complex decisions.
Recognition primed decision (RPD) model	A combination of rational and intuitive DM where a decision maker considers only one option instead of weighing all. Steps include: – Identify the problem, including all characteristics, expectations and business goals. – Think through the plan and conduct a mental simulation to see if it works and what modifications might be needed. – If the plan seems OK, make the final decision and implement it.	Applicable to situations that call for an immediate decision	Success depends highly on the decision maker's experience and expertise.

TABLE 2. INDIVIDUAL DECISION MAKING STRUCTURES AND PROCESSES

Decision Making Style	Description	Advantages	Disadvantages
Creative / brainstorming	Decision makers gather information and insights regarding the issue and develop preliminary solutions. Then, they enter an incubation period where they do not actively think about the alternatives. They allow their unconscious to take over in hopes that it will lead them to a realization and answer, which they can then test and finalize.	May provide solutions that are not immediately obvious.	Can be time-consuming for complex decisions.

3.5. ORGANIZATIONAL STRUCTURES FOR DECISION MAKING

From an organizational perspective, two main organizational structures impact DM processes - centralized and decentralized. In a centralized structure, DM authority is concentrated at the top levels of the organization hierarchy (essentially the "Top-Down" process described in Table 2). In a decentralized structure, DM is distributed across different levels and units of the organization (through various methods, such as the "consensus" and "distributed" processes described in Table 2). Table 3 summarizes the two approaches.

TABLE 3. ORGANIZATIONAL STRUCTURES FOR DECISION MAKING

Organizational structure	Description	Advantages	Disadvantages
Centralized	Senior managers/executives make decisions at the top. Information flows upwards from lower levels to top decision makers	Allows for faster and more coordinated decisions across the organization.	Senior managers/executives may lack proximity and context for optimal decisions.
Decentralized	DM is distributed across different levels and units of the organization.	Decision authority is delegated to managers/teams closer to the work. Enables faster decisions by those with relevant expertise and context. Promotes employee empowerment and engagement in DM.	Requires strong DM capabilities at all levels. Can lead to inconsistencies if decisions are not properly coordinated.

The appropriate structure depends on the organization's goals, culture, and the nature of the decisions required. Centralized structures are suitable for decisions requiring tight coordination, consistency, and speed across the organization. Decentralized structures work better when flexibility, innovation, and context-specific choices are needed.

Many organizations adopt a hybrid approach, centralizing strategic decisions while decentralizing operational decisions to leverage the benefits of both models.

Effective managers align the organizational structure with DM needs, empower teams appropriately, and foster transparency and collaboration for optimal outcomes.

In new build projects, it is important to have clarity on the endpoint structure and the milestones on the journey so that the organization's goals, culture, and the nature of the decisions required at each stage of the lifecycle of NPPs are effectively linked.

4. RISK INFORMED DECISION MAKING

4.1. RISK AND RISK ASSESSMENT

DM in NPP owner/operator organizations involves analysing and transmitting accurate risk information to decision makers. One of the Nuclear Energy Basic Principles is that "the use of nuclear energy should provide benefits that outweigh the associated costs and risks" [18], so risk assessment is a key part of DM. Many tools have been developed in a nuclear facility context to assess risk. These include probabilistic safety assessment (PSA) technologies, failure modes and effects analysis (FMEA), hazard and operability analysis (HAZOP), human reliability analysis (HRA) methods, external hazards analysis (seismic, flooding, etc.) and fire risk analysis. Appendix III describes each of these.

4.2. RISK INFORMED DECISION MAKING

Risk informed decision making (RIDM) is a structured approach that incorporates risk assessment insights into the DM process for complex systems or operations. The following sub-sections discuss RIDM from several international perspectives.

4.2.1. IAEA's Integrated Risk Informed Decision Making framework

IAEA's integrated risk informed decision making (IRIDM) publication TECDOC-1909 [9] defines a DM process that applies primarily to safety issues but applies (per the document's Section 2.3) to a wide variety of topics ranging from design, licensing, operation, maintenance, operator training, and modifications.

Risk considerations are explicitly addressed, and the insights from the deterministic and probabilistic analyses are combined with other requirements (from the regulatory body or the operating organization) and considerations (cost-benefit, sound engineering practices, etc.).

Integrated risk informed decision making systematically and holistically considers many relevant factors and brings together different aspects and considerations into a single framework, including key elements of the process (See Fig. 1. Key elements of the integrated risk informed decision making process in Ref. [19]), establishing a multi-disciplinary team of specialists, gathering necessary information and providing for expert review of uncertainties.

4.2.2. INPO's Principles for Excellence in Integrated Risk Management

The Institute of Nuclear Power Operations' Principles for Excellence in integrated risk management (INPO 15-011 [20]) extends beyond PSA techniques. It provides a broader set of insights to be considered in DM, indicating that effective integrated risk management involves understanding the interrelationships and managing the strategies for enterprise risk, operational risk and project risk. Its principles centre around culture, individual responsibility, governance, risk determination and DM, risk mitigation, communication of risk, and self-evaluation and learning.

4.2.3. Risk informed decision making at US NPPs

The B. John Garrick Institute and Japanese Nuclear Risk Research Centre (NRRC) belonging to Central Research Institute for Electric Power Industry (CRIEPI) published a report [21] on RIDM in the USA, together with an overview of the history of PSA and nuclear regulation in the USA. The applications include 20 cases ranging from outage risk management, training, in-service inspection, and communication.

The United States Nuclear Regulatory Commission (USNRC) provides a guide [22] on using PSA in risk informed decisions on plant-specific changes to the licensing basis, in which acceptance guidelines are given for a possible incremental change in core damage frequency (CDF) and large early release frequency (LERF) triggered by plant configuration and procedural changes. (See Fig. 4. Acceptance guideline for core damage frequency in Ref. [22] and Fig. 5. Acceptance guideline for large early release frequency in Ref. [22])

The current key US industry-wide initiatives to improve DM using a risk informed approach are described in Appendix VII.

5. ORGANIZATIONAL DECISION MAKING

There needs to be a constant flow of decisions in an organization at all levels, from the boardroom to the plant floor. DM processes need to ensure that DM functions effectively throughout the organization, leading to high performance. The desired outcome is effectiveness and efficiency in DM, focusing on resilience and adaptability.

This section examines the various levels of an organization and their role in DM, plus some mechanisms and attributes that can be implemented to enhance DM.

5.1. CORPORATE GOVERNANCE AND DECISION MAKING

5.1.1. Board of Directors and senior management roles

The IAEA defines "senior management" as "the person or persons who direct, control and assess an organization at the highest level" [23]. These individuals fall into two categories: those responsible for the organization's day-to-day operation and those in an oversight and advisory role, often as members of a Board of Directors or similar oversight body. Such an oversight body ensures that an organization operates lawfully and in the interests of its shareholders and other stakeholders (e.g., its employees).

In a nuclear context, the Board needs to specifically oversee the organization's nuclear safety and security risks, as well as those risks common in any industry. Sufficient nuclear knowledge and experience need to exist within a Board's collective membership to enable effective DM, including understanding the special and unique nature of nuclear plant operations.

On the other hand, even board members with limited expertise in nuclear power generation have to acquire fundamental knowledge about it and recognize the significant risks involved. They need to understand how decisions made by the nuclear division can impact the overall risk to the company and ensure that these decisions do not lead to catastrophic outcomes. They need to question and actively challenge these decisions to curb the excessive autonomy of the nuclear division. In this way, the Board facilitates ongoing risk monitoring and establishes policies that allow the organization to measure, minimize and manage project, operational and enterprise risks.

The Boards, with the support of other senior management, are typically responsible for:

- Ensuring a nuclear safety and security policy is established;
- Communicating a clear, unambiguous message that nuclear safety is the company's highest priority;
- Ensuring that the nuclear component of the company has clear strategic direction, with a clear vision, mission, values and strategic goals, with supporting operational plans, which places nuclear safety and security paramount;
- Ensuring clear guidance is produced related to environmental, social and governance (ESG) issues (see Section 6.9.3) to aid in DM (e.g., by including direction in the organization's mission, vision, values, policy statements, authority registers, training and routine corporate communications);
- Ensuring clear standards for ethical conduct are enshrined with company policies and values;

- Using diverse sources of information – such as performance indicators, self-assessments and independent assessments, quality assurance reports, and corrective action trends – to identify and allow early correction of adverse performance of the organization;
- Ensuring operational risks and long-term strategies associated with nuclear plant performance are measured, minimized and managed;
- Establishing independent oversight methods related to nuclear safety and security that report directly to the Board.
- Receiving objective input from independent oversight regarding site and corporate performance compared to industry performance standards, focusing on nuclear safety;
- Periodically visiting the nuclear stations to communicate directly with personnel on-site and to observe activities and plant conditions;
- Providing oversight of the organization's talent management strategy and ensuring it supports the business strategy.

Expert committees may support Boards that provide advice and inform key decisions regarding engineering, nuclear operational performance, training, and other key areas.

The board structure will need to adapt and change throughout the lifecycle of NPP. The board structure during construction will be different from that during operation. However, the makeup of the board has to ensure there is effective representation to cover the whole lifecycle at all times.

While the requirements for senior management focusing on safety are described in IAEA Safety Standard on Leadership and Management for Safety, GSR Part 2 [24], IAEA Safety Guide GS-G-3.5 [25] Section 2.10 further describes senior managers' DM roles, stating that:

> "Senior management should establish and promote a set of principles to be used in decision making and promoting safety conscious behaviour. Examples of such principles used in some organizations are as follows:
>
> (a) Everyone has an impact on safety.
> (b) Managers and leaders must demonstrate their commitment to safety.
> (c) Trust and open communication permeate the organization.
> (d) Decision making reflects putting safety first.
> (e) Nuclear technology is recognized as having unique safety implications.
> (f) A questioning attitude is fostered.
> (g) Organizational learning is encouraged.
> (h) Personnel training is encouraged.
> (i) A proactive approach to safety is taken.
> (j) Safety is constantly under review."

Senior management encounters a wide range of issues beyond safety issues and relevant DMs in NPP operating organizations. In this context, the role of senior management in DM would include but not be limited to:

- Ensuring that the DM rule (process, decision authority) is understood and adhered to by the people working for the NPP operating organization.

- Ensuring that considerations are given to safety, security, budget, societal implications, etc. and that systematic biases are avoided in DM.
- Monitoring execution of the corporate nuclear strategic decision and intervening, when needed, to ensure strategic goals are obtained and the results are sustainable. More specifically, observing the limitations of the decision executing team objectively and deciding on corrective actions such as:
  - Arranging an opportunity to listen to and/or have support from experts in different disciplines (and add them to the team) or outside of the organization for out-of-the-box thoughts;
  - Securing additional resources (i.e., human, financial, others);
  - Postponing the decision until enough information is collected or sufficient resources are provided.
- Fostering informed DM, with the understanding that nuclear technology is special and unique and making sure that a timely decision is made before it is too late;
- Making sure that high priority is placed in DM at all the levels on:
  - Developing current and future nuclear leaders and the effectiveness of corporate and site leadership teams. (See Section 5.3)
  - Promoting collaboration among the support groups and non-nuclear business units to drive a coordinated approach for supporting nuclear stations;
  - Building and sustaining trust with internal and external stakeholders through timely, accurate, and transparent flow of information.

5.1.2. Establishing clear lines of decision authority and accountability

Senior management needs to define authority levels ("who decides what") and the required knowledge and experience for organizational DM as part of implementing a management system. Establishing organizational role documents that require sufficient nuclear experience for key roles will help facilitate good DM.

Signing authorities can be defined for such areas as purchasing or contracting limits, project approvals, change notice approvals, write-offs, business expenses and similar financial matters, human resources decisions (hiring, termination, salaries and organization structure), information management, strategic planning and for plant operations and maintenance. Clear authority-level documents and management system governance have to clarify whether personnel at any level of the organization can make decisions in a defined area.

WANO's Traits of a Healthy Nuclear Safety Culture [26] indicates that a health safety culture trait related to accountability for decisions is that "Individual or single-point accountability is maintained for nuclear safety decisions."

Problems can arise if DM is too high in the hierarchy, resulting in slowness and distraction of senior management, or too low, leading to misalignment between departments or decisions not in alignment with organizational goals. Guiding broader organizational performance (through mission, vision and values) and offering strategic information to senior managers can help ensure alignment, high performance and organizational resilience. Independent oversight and reviews of decisions made at lower levels may identify areas where there are organizational misalignments.

5.1.3. Checks and balances in organizational decision making

Checks and balances help ensure that the appropriate individuals or bodies make decisions with the requisite expertise and authority. Such features promote accountability, prevent unauthorized or unilateral DM, ensure ethical behaviour, and help identify and mitigate potential risks, errors, or oversights arising from decisions made without proper oversight or review.

Checks and balances can include:

- Independent reviews and approvals for significant decisions or transactions;
- Oversight mechanisms like safety committees or advisory boards;
- Segregation of duties and authorities across different roles/functions.

Some examples common in nuclear include:

- Policies that require certain decisions to be approved by executive management or the board;
- Processes and procedures that mandate independent verification, peer reviews, or cross-functional approvals before finalizing decisions;
- Organizational structures and roles that segregate DM authorities across different functions (e.g., operations, safety, quality assurance) to provide checks and balances. An example would be having approval for continued unit operation rest with the design/engineering authority as opposed to a production manager;
- Establishing audits or formal reviews of important decisions.

5.1.3.1.   Design authority

The operating organization is responsible for controlling plant configuration and thus needs a process to maintain and control the plant's design basis and plant configuration. The entity that oversees these functions is the design authority.

The design authority (DA) approves and documents the design basis, licensing basis, and plant configuration and formally approves all design changes. The DA normally is the top decision maker for these areas, with veto power. A strong DA can thus act as an effective check and balance to any DM related to these areas.

Other senior management need to respect and reinforce the authority of the DA and the importance of understanding, respecting, promoting and reinforcing technical considerations and conservatism in decisions that reflect the need to operate and maintain the plant within the design, safety and licensing requirements. They also need to ensure an appropriate balance between technical conservatism, plant operating margins, and business needs for matters that affect plant reliability.

INSAG-19 [7] recommends the implementation of a DA function. IAEA SSR-2/1 (Rev. 1) [27] transferred the recommendation into a safety requirement stating that "each operating organization shall establish a formal system for ensuring the continuing safety of the plant design throughout the lifetime of the nuclear power plant." Accordingly, the entity providing the DA function is included within the operating organization's management system. The World Nuclear Association (WNA) has compiled a special report [28] describing the implementation of the DA.

When performing its function, the DA needs to remain independent from the responsible designers. However, before a decision is made on a modification, the DA needs to meet the Responsible Designer to understand the design basis that will form the constraints within which the modification needs to conform (See Section 7.3.1 in Ref. [28]). In small companies, it can be practical to contract design work to external contractors such as those from the plant vendor. However, the DA function has to always be within the licensee's organization.

A model of DA, as implemented in France, is provided in Section VII.1.2 of Appendix VII.

5.1.3.2.   Independent oversight

Independent oversight needs to scrutinize whether the DM procedures are correctly in place and adhered to by the people working in an NPP operating organization. Independent oversight may challenge, for example, the decision being made to restart the plant after an outage by reviewing the relevant DM process and the evidence on which the decision is based.

Leaders and decision makers at various organizational levels need to have direct access to the results of independent oversight.

Guideline, GL 2018-01, on independent nuclear safety oversight for operating organizations, was published by WANO and IAEA in 2018 [29].

5.1.3.3.   Cross-functional decision making teams and committees

Cross-functional teams, with representatives from different parts of an organization, provide diverse perspectives, knowledge, and expertise by bringing together individuals from other functional areas and backgrounds. This diversity allows for more comprehensive analysis, creative problem-solving, and well-rounded decisions considering various aspects and potential impacts. Benefits can include:

- Improved coordination and alignment (information is better shared across the organization, leading to more cohesive and aligned DM);
- Holistic views and systematic thinking (teams take a look at the entire organization and its processes, enabling the review of interdependencies);
- Increased buy-in (cross-functional teams foster a sense of ownership and commitment to the decisions made);
- Enhanced innovation and creativity (diverse backgrounds and perspectives stimulate creative thinking and innovative solutions);
- Improved decision quality (with a broader range of information, expertise, and perspectives, cross-functional teams can make more informed and higher-quality decisions).

Some challenges for cross-functional teams can include:

- Need for team members to support the best solution for the organization and not the part of the organization that they represent (need to break down silos);
- Need to clarify accountabilities and responsibilities among team members;
- Differing communication methods or processes among team members; differing power dynamics;
- Lack of trust and respect (especially when team members have not worked together).

A common use of cross-functional DM for difficult operational decisions with significant uncertainty is operational decision making (ODM) or addressing operational decision making issues (ODMIs). The ODM process is a participatory DM technique (see Section 7.4) by a committee of senior facility managers that includes six steps: prompt recognition, consistent review, rigorous evaluation, consequence-biased decision, effective implementation and periodic assessment. Refer to WANO Guideline WANO GL-2002-01 [3] for details on this process. Cross functional teams may also be deployed to address complex technical or operability concerns. Nuclear Energy Institute publication NEI 18-03 [30] describes a formal process for technical operability evaluations.

5.1.4. Formal decision making processes and management systems

Formal DM processes and approval workflows provide a structured and consistent approach to DM, ensuring that important decisions are made methodically and relevant factors are properly considered. They encourage rational DM, discourage DM based on biases or heuristics, and contribute to ensuring that a clear basis for a decision is agreed upon, documented, and understood. Documentation can include the decisions made, how alternatives were considered, and the rationale for the decision. Formal processes promote accountability and transparency by defining clear roles, responsibilities, and approval steps, thus improving consistency, aiding problem-solving and facilitating later review [31].

An organization's management system provides the structure for these formal DM processes. Management systems, implemented using good project, programme and change management principles, document an organization's mission, vision, values and strategy, organization structures, and roles and responsibilities. They also provide performance measurement systems, feedback loops, audits, and post-implementation reviews to evaluate organizational DM and capture lessons learned.

Analysing past decisions that were found to be inappropriate can yield valuable lessons. Independent, objective reviews of such decisions can uncover organizational structural or cultural issues that may be sensitive or not immediately obvious. Figure 5. in IAEA TECDOC-2038 [2] provides an example of such an assessment of why bad decisions were made.

Management system documents such as policies, standards, procedures, checklists and forms provide guidance and instructions to personnel on implementing decisions within their organizations per the requirements imposed on them. Mandatory review checklists, procedures, and defined workflows within enterprise IT systems can help improve decision quality and ensure compliance. Some examples of checklists are provided in Appendix V. The USNRC analysed the positive impact on human reliability of written procedures in a 2016 report [32], which indicated that "procedure-based operation eliminates many challenging and uncertain aspects in DM: the team structure and working protocols in the control room do not demand many cognitive aspects of communication and coordination activities."

Nuclear management system processes need to define who is responsible for making decisions and approving actions, thus reducing ambiguity and potential conflicts. IAEA Safety Guide GS-G-3.5 [25] Section 2.16 reads, in part, as follows, indicating the importance of defining DM authorities in nuclear organizations:

> "Managers should establish the authorities and decision making powers for all positions in the organization. These powers should be exercised, and there should be clear lines of authority for safety-related matters. Accountability means that all individuals should know their specific assigned tasks (i.e. what they have to accomplish and by when, and how to recognize good results); if individuals are unable to execute their assigned tasks as expected, they should report this to their supervisors."

IAEA requirements and guidance related to management systems are included in safety standards such as GSR Part 2 [24], GS-G-3.1 [33], GS-G-3.3 [34], GS-G-3.4 [35], GS-G-3.5 [25], TS-G-1.4 [36] and guidance publications such as Nuclear Energy Series No. NG-T-1.3 [37]. IAEA Nuclear Energy Series publication No. NG-T-1.6 [12] provides guidance on project management of nuclear power plant projects, and TECDOC-1226 [13] discusses managing change in nuclear utilities. An online Management System Network of Excellence[3] may also be consulted.

5.2. KNOWLEDGE MANAGEMENT

Knowledge, including relevant data, information, and experience, provides the DM processes within nuclear organizations with the correct and needed resources. Knowledge management (KM)[4] processes that capture expert knowledge and lessons learned that are integrated into the information-gathering, structured DM, and organizational learning processes described earlier in this section can improve decision quality dramatically. Knowledge management prevents the loss of critical expertise over time by codifying knowledge from experienced employees, lessons learned, past decisions, and external sources. This organizational memory aids in avoiding repeated mistakes and retaining DM capabilities.

External sources of knowledge that are useful for KM and DM include:

- Technology-specific Owners Groups, communities of practice and conferences for specific subject matters;
- National laboratories and research & development institutions;
- International nuclear organizations such as the IAEA, OECD-NEA, WANO, INPO and various regulatory forums;
- External subject matter experts, where specialized knowledge is unavailable internally to an organization.

IAEA Nuclear Energy Series publication No. NG-G-6.1 [38] provides guidance on KM strategies and approaches in nuclear energy organizations and facilities. Numerous other IAEA KM resources are available, including Refs. [39–46] and a KM assist visit service [47]. An AI

---

[3] See https://nucleus.iaea.org/sites/connect/MSNpublic/SitePages/Home.aspx

[4] Knowledge management is a broad topic and cannot be fully covered in this publication. In this publication only the context of decision making is discussed.

example from EDF that helps detect early component problems is described in Appendix VII.2.4.

5.3. LEADERSHIP AND CULTURE

5.3.1. Leadership

Senior managers benefit from possessing and demonstrating the following leadership attributes/skills to fulfil their expected roles and responsibilities in DM, besides the personal qualities of integrity, empathy, dedication, etc.:

–   Showing a vision for the future, briefing how they are linked with individual decisions, and setting an example for desired performance per the organization's goals to inspire employees;
–   Ensuring the organization's vision, mission, and goals are embedded within and supported by the entire management team in the course of the DM process;
–   Capturing and sharing with employees the status-quo (the current situation/context the organization faces) for "sense-making";
–   Readily adapting DM strategies to different contexts and needs [48];
–   Being directive when the situation requires such action;
–   Demonstrating emotional intelligence to lead employees, balancing emotion with reason, and listening to others;
–   Communicating, empowering and involving employees in DM to get a wide choice of alternative ideas or diverse options and to encourage their debates and deliberation for effective DM.

Industry guidance documents from INPO (15-005 Rev. 1, especially section LE.8 on DM and coping with risk [49]) and WANO PL-2019-01 [50] also provide insights into leadership and DM skills.

Key elements in such decisions include:

–   Does the decision align with the organization's mission, vision and objectives?
–   What is the potential impact of the decision or a no decision (i.e., delaying the decision)?
–   Does the decision, if impacting nuclear safety, emphasize a prudent, conservative choice over being simply allowable?
–   Is it clear what has been decided, why and how it will be implemented?
–   Does the decision address the root cause of the issue (rather than a symptom)?

The IAEA Safety Standard on Leadership and Management for Safety, GSR Part 2 [24], emphasizes the key role of senior management in this effort. Appendix I in the IAEA TECDOC-2038 [2] provides an example of leadership and effective DM demonstrated by the head of Fukushima Daini Nuclear Power Station during its emergency response in March 2011.

5.3.2. Organizational culture and traits

A decision made by an organization reflects its culture or traits, in particular, the health of its nuclear safety culture. WANO's Traits of a Healthy Nuclear Safety Culture [26] identifies DM as an important area for consideration, indicating that the following has to be apparent in a healthy nuclear safety culture:

"Decisions that support or affect nuclear safety are systematic, rigorous and thorough. Operators are vested with the authority and understand the expectation, when faced with unexpected or uncertain conditions, to place the plant in a safe condition. Senior leaders support and reinforce such conservative decisions."

Opportunities for correct decisions would increase as decisions are made by recognizing these cultures or traits and then avoiding possible traps attributable to them.

Some negative examples may be:

- When facing difficult decisions, decision makers may defer a decision until the situation changes or new information emerges. While this can be beneficial in some cases, it may also result in missed opportunities or worsen the problem;
- Peer pressure and groupthink can deprive the organization of valuable discussion[5] and scrutiny of proposed solutions;
- Sectionalism without lateral integration leads to a wrong decision by ignoring potential issues outside of one's area of expertise;
- Other culture/traits or tendencies, such as below, may also lead to poor decisions:
  - Insufficient learning culture;
  - Unclear core values and priorities;
  - Poor understanding of scientific and technological basics;
  - Over-confidence and lack of respect for experts;
  - Lack of humility in listening to diverse views;
  - Excessive deference to seniority, resulting in withheld dissenting opinions;
  - Short-sighted and narrow-minded thinking.

Positive examples may be:

- Inviting alternative views from others for diverse perspectives and options and installing a DM rule that assigns a "devil's advocate" to present dissenting or critical views and for vitalizing discussion;
- Fostering a culture of continuous improvement that scrutinizes decisions afterwards to determine if they were correct and yielded expected results. If not, the organization analyses why ineffective decisions were made.

As discussed in Section 3.2, it is recommended that a decision maker, an oversight group or third-party experts assess why good and/or wrong decisions were made. This approach would help in understanding organizational culture and trait issues.

---

[5] Amazon's "Golden Silence or Silent Start (before the meeting) meeting method" is an example to provide space for diverse thinking, deepening understanding of the issue and reasoning of strength/weakness of the proposal before the DM meeting. [51]

5.3.3. Developing competency and capability in decision making

Managers at all levels need to possess the necessary skills to make effective decisions and to provide oversight and support to subordinates making decisions within their span of control. They need to understand and be always aware of the basics of DM discussed in Section 3, including the structured DM process, human elements (heuristics, bias, trap), DM style and organizational structures for DM. While people gain part of these skills over time, formal training programs can enhance individuals' DM skills and shape the company's culture as well. These programs range from explaining the factors that influence DM as described in the IAEA SSG-75 Safety Guide [52] to scenario-based training to prepare managers for difficult decisions they may encounter. Like any training, a combination of principles, practice and knowledge of pitfalls can enhance managers' abilities to make effective decisions. Such training is not one size fits all, and managers have to be trained to adapt to different circumstances and adopt different DM styles and strategies to make optimal decisions (e.g., slow vs fast, directive vs collaborative, etc.). Adapting to different contexts is one of the key attributes for resilient operations highlighted in the IAEA NE Series publication on "Strengthening Resilience in the Nuclear Organization," currently being drafted. Training on RIDM (see Section 4.2) is especially important for nuclear organizations.

Additional techniques for building DM skills include:

- Using DM exercises like case studies, role-playing, and simulations to practice without direct consequences;
- Identify a mentor, such as someone more senior in the organization, shadow them, and discuss key decisions in the workplace;
- Gather regular 360° feedback from the organization, including superiors and peers. This feedback can reflect on recent decisions or serve as input in the DM process. The feedback needs to be candid, reflecting success and failure and future improvement opportunities;
- Practice increasing leadership responsibilities, especially when emerging as a newer manager. This leadership can be formal regarding work assignments and include volunteer and mentoring opportunities beyond current job roles;
- Practice communication skills such as active listening to discussions with those who hold opposing viewpoints. Do not surround yourself with those who agree with your views, but seek out those who can expand your views;
- Build your knowledge by staying current on technical developments related to nuclear energy. Likewise, read articles and books on DM processes to refresh and expand your DM skills.

6. OWNER/OPERATOR DECISION DOMAINS AND KEY CONSIDERATIONS

This section describes the various domains within the nuclear industry where challenging or tough decisions are made and some key considerations that senior managers need to address to facilitate effective strategic and tactical DM.

6.1. STRATEGIC PLANNING AND DECISION MAKING

Senior managers at a nuclear facility or its parent organization face regular strategic planning and investment decisions in the following areas:

- Approvals of major financial investments in the facility, including those for:
  - Initial siting, design, construction and commissioning;
  - Investments during the operating life to increase reliability or power output, improve nuclear safety or increase safety margins, address changes in regulatory requirements, reduce occupational or public radiation exposure, reduce operating and maintenance costs, manage supply chain risks, or reduce industrial safety risks;
  - Investments to preserve or extend the facility's life, including those associated with long-term operation (LTO).
- Approvals for human resource increases (hiring, contracting) or decreases (layoffs, retirement packages, contract cancellations) and investments in training and skills development for the workforce.

Such decisions are compared to other opportunities for investment by the organization, which may include alternate electrical capacity investments (nuclear or non-nuclear).

To address these strategic issues and tactical planning, NPP owner/operator organizations typically develop processes and frameworks that provide the necessary information to support effective DM. A typical framework may contain some or all of the following elements:

- Integrated resource planning (IRP)

  IRP evaluates future demand, assesses existing and potential new generation sources (including advanced nuclear technologies like small modular reactors (SMRs)), and develops optimal resource portfolios over a long-term horizon (20-30 years). IRP has to consider load forecasts, fuel supply, system reliability, environmental impacts, economics and risk.

- Nuclear technology road mapping

  Roadmaps assess the maturity, costs, timelines, and deployment potential of advanced nuclear reactor designs like SMRs, as well as improvements and research and development associated with in-service designs. These roadmaps allow organizations to decide when to pursue new nuclear investments.

- Regulatory engagement and harmonization

  Close coordination with nuclear regulators on reactor licensing issues and regulatory harmonization is crucial for successful long-term planning.

– Stakeholder involvement

Structured stakeholder engagement activities help organizations understand the perspectives of governments, local communities, environmental groups, and others. This engagement builds public confidence and identifies issues to be addressed.

– Organizational agility

With rapid technological changes, organizations need to cultivate the agility to adapt quickly. This agility includes developing skilled workforce pipelines, embracing modern project management approaches and advanced IT solutions, and fostering a culture of innovation.

– Financial modelling and risk assessment

Financial modelling and risk assessment are crucial in evaluating strategic investments for nuclear utilities. Financial models help analyse proposed investments' economic viability and profitability over their lifecycle. Common techniques include net present value (NPV) analysis, internal rate of return (IRR) and payback period.

The IAEA provides resources for financial modelling and contracting for NPP projects. These resources include various software models (e.g., FINPLAN and others, see Ref. [53]), a Nuclear Contracting Toolkit [54], publications [55–58], and assist missions. The World Nuclear Association (WNA) provides a guide on structuring nuclear projects [59].

6.2. PLANT OPERATION AND MAINTENANCE

Some strategic and tactical decisions regularly encountered related to plant operation and maintenance include:

– Multi-year outage and on-line maintenance planning;
– Approaches for degraded plant conditions or new information where clear instructions or guidance do not exist or where uncertainty exists;
– Long-term vision on equipment repair/replacement and lifecycle management decisions;
– Balancing production needs for grid stability with plant safety and regulatory requirements;
– Communications and interactions with the local community regarding plant events.

The resolution of these complex plant management issues can be supported by the above-discussed elements in Section 6.1 and also the cross-functional DM process discussed in Section 5.1.3.3.

6.3. PLANT DESIGN IN NEW NUCLEAR BUILD

A huge number of decisions need to be made to complete new plant design at various levels, from the top-level concept to detailed specifications. In order to effectively make plant design sustainably reliable and to enable efficient operation and maintenance, some or all of the following elements need to be considered:

– Recognition of reversibility of decision

It is important to understand that some decisions are reversible, and others are irreversible in plant design work. In case it is recognized that the decision is reversible, we can introduce flexibility into the decision process (See Step 8 in Fig. 2. in Section 3.1), take a more nuanced approach and increase the possibility of reaching a more appropriate decision. On the other hand, if it is recognized that the decision is irreversible, it is necessary to pay more attention to that decision, considering its potential long-term impact. More scrutiny might be required, and more resources need to be allocated with priority.

– Flexibility and margin in the new designs

NPP owner/operating organizations can not avoid considering possible unforeseen natural hazards, climate change impacts, unexpected material degradation, cyber security threats, socio-political issues, etc. Although preparedness for completely unknown threats is theoretically impossible, flexibility and margin in the design may help cope with completely new threats. Such initiatives would include:

- Introduce the concept of 'do not stop thinking about anything,' specifically, develop procedures for responding to any events with very low probability and/or large uncertainties beyond design basis scenario and implement more diversified initiatives/strategies to increase the possibility of managing them;
- Conduct regular 'what-if' scenario exercises;
- Foster cross-disciplinary collaboration across organizations and industries;
- Utilize advanced simulation and modelling with cutting-edge simulation tools, etc.

– Enhancement of discussion and communication between designers and operators

Close collaboration and communication between designers and operators at the initial stage of plant design are essential. It is especially important to convey requirements from operators to designers and discuss constraints. That allows for better integration of design and operational considerations, facilitates informed DM and improves overall design quality.

– Planning fallacy

Designers tend to be trapped by optimism bias when it comes to the cost/price of a newly designed plant. The cases of EPR and AP1000 cost overrun are due to this optimism by a) not considering the loss of experiences in construction projects in the last three decades in the US/Europe and b) the start of construction before detail design is complete (thus, repeated design changes). This fallacy significantly damages the perception of the economics of nuclear power among the public and the financial world.

6.4. SUPPLY CHAIN ISSUES

Strategic decisions for NPP owner/operator organizations related to the supply chain include:

- Investments in supporting suppliers to maintain the capability to provide critical components, spare parts or specialized skilled capabilities (e.g., testing, inspection, fabrication, technical support, R&D). These investments may include initiatives to ensure adequate human resources and skills to support supply chain work;
- Investment in supporting quality improvement and training of critical supplier personnel (e.g., in topics such as human performance or leadership that are critical to the nuclear industry);
- Special production runs of soon-to-be-obsolete components;
- Establishing company-wide standards for technology selection or certain equipment or component types to minimize warehousing, maintenance and supplier management costs, considering the needs of existing and proposed new plants;
- Establishing master service or partnership agreements with key engineering, trades or component suppliers to maintain minimum volumes of work (to maintain supplier capability) and facilitate dialogue, minimize warehouse costs, improve performance and share information;
- Reduction in the number of suppliers used (to minimize transaction costs) or an increase in the number (to address performance issues, encourage competition, encourage procurement from designated groups, or address industry expansion efforts);
- "Make or buy" decisions regarding what capabilities to keep in-house versus those to be contracted out and establishing an informed customer capability. This decision is impacted by the degree to which the original technology vendor is relied upon versus third-party engineering companies or in-house engineering capability.
- "Repair or replace" decisions aligned with maintenance and warehousing strategies to minimize the cost of keeping certain equipment in service;
- The degree to which environmental, social and governance (ESG), climate change considerations and similar initiatives (see Section 6.7) are weighted in purchasing decisions.

NPP owner/operator organizations need to pay close attention to these supply chain issues in support of reliable operation. Due to the long lifetime of nuclear facilities, obsolescence of components or loss of skills over time is a threat. A region can move from being heavily new build-influenced, with many suppliers, to more operations-influenced, with smaller numbers of suppliers. Globalization of supply chains can make supplier oversight difficult and add risk to NPP operation and management. Having a clear overview of the full supply chain is essential.

Additionally, government-owned or influenced organizations may need to purchase from local suppliers or meet other government-mandated procurement objectives. These objectives can impact supplier availability, quality and costs.

Supply chain companies need time to respond to industry changes. They will not invest in new capabilities or maintain current ones unless there is a compelling business case. Regular communication between NPP owner/operator organizations and supply chain participants at various levels is needed to understand each organization's planning assumptions. A spirit of partnership needs to be followed rather than a pure contracting approach.

Engaged suppliers, including nuclear technology vendors, can assist in DM related to certain technical issues. A supplier's knowledge and related information (e.g., plans, initial design documents, technical knowledge) may be fundamental to making informed decisions concerning troubleshooting, operability, repair or replacement issues.

IAEA guidance related to vendor management and the supply chain is available in Nuclear Energy Series publications NP-T-3.21 on procurement engineering and supply chain management [60], NG-T-3.4 on industrial involvement [61], and a Nuclear Contracting Toolkit [54]. Appendix VII.1.3 provides some supply chain examples from EDF.

6.5. REGULATORY COMPLIANCE

Nuclear facilities need to pay close attention to regulatory compliance. Functions and decisions related to this area include:

— Interpreting and applying nuclear safety regulations and requirements;
— Interacting with nuclear regulators and oversight bodies;
— Reporting and notifications of events;
— Understanding relevant risks and benefits of regulations and requirements;
— Communicating regulatory changes and interpretations to operational staff.

The interactions with regulatory bodies are particularly important. Unneeded or inefficient legacy rules and requirements can become costly for nuclear operators. However, proactive discussions between operators and regulators can help define common approaches to complex or new issues or opportunities, including new designs, modifications, technologies, changes in public sentiment and stakeholder acceptance. Engagement with international initiatives to develop and harmonize safety standards can provide tangible benefits.

6.6. GOVERNMENT POLICY

Governmental policies invariably impact NPP owner/operating organizations. Whether directly controlled as part of a Ministry of Energy or a government-owned subsidiary, or indirectly as a private company reporting to its shareholders, such organizations are impacted by government energy policies related to electricity production technology mix, market design, regulatory regimes, applicable laws and standards, public consultation requirements, and project approval requirements. In the early stages of nuclear power programmes, clear government support for nuclear power development is essential for progressing the programme and addressing the many infrastructure issues described in the IAEA milestones approach [62].

Proactive discussions with government officials are needed to ensure that decision makers in those organizations have the information to make high-quality decisions related to the nuclear power industry.

6.7. NUCLEAR SECURITY AND SAFEGUARDS

Nuclear security and safeguards are critical components of responsible nuclear energy management. While these essential functions may not directly contribute to electricity production revenue, they are vital in ensuring public safety, maintaining international trust, and upholding the integrity of nuclear operations. Some strategic issues requiring decisions include:

- Decisions to outsource or staff internal security services, including the emergency response force;
- Maintaining compliance with evolving nuclear security and safeguards requirements;
- Evaluation of potential changes in the design basis threat based on regional and global conditions;
- Level of investment in addressing computer security/cybersecurity risks;
- Level of concern and oversight of potential insider threats.

6.8. NUCLEAR WASTE MANAGEMENT

National strategies, fuel cycle considerations and regulations impact nuclear waste management decisions. Siting decisions related to waste storage, reprocessing, and disposal facilities have high stakeholder interest and political overtones. Some critical decisions include:

- Spent nuclear fuel storage and disposal pathways;
- Radioactive waste classification and treatment methods;
- Decommissioning strategies for retired NPPs;
- Transportation and interim storage facility decisions.

6.9. OTHER KEY CONSIDERATIONS

6.9.1. Economics

Economic and financial considerations invariably influence decisions. Detailed economic analysis may include modelling to assess a decision's direct and indirect economic effects, aiming to ensure equity in the distribution of benefits and burdens.

Understanding the purposes of economic incentives and government policies is also very important for making decisions. This work may include analysing tax structures, financing terms, market structures, regulatory frameworks, policies to assist targeted groups or provide local employment, and government subsidies that can influence the organization's operations. The organization needs to actively participate in policy advocacy to shape an enabling environment that supports its strategic planning and contributes to its DM.

6.9.2. Morals and ethics

It is crucial to incorporate ethical principles and values into the DM process beyond just legal and regulatory compliance. In alignment with an organization's values and code of conduct, ethical DM can promote trust, fairness, accountability, and social responsibility.

Ethical principles that would guide DM include honesty, integrity, respect for others, justice, fairness, beneficence (doing good), and non-maleficence (avoiding harm). The IAEA's Nuclear Energy Basic Principles [18] are aligned with these and provide additions such as security, non-proliferation, long-term commitment, resource efficiency, and continual improvement.

Common challenges that can hinder ethical DM include:

- Conflicting values or interests;
- Ambiguity or uncertainty about the consequences of a decision;

- Pressure to prioritize short-term gains over long-term sustainability;
- Organizational cultures that do not prioritize ethics;
- Cognitive biases and rationalizations leading to unethical choices.

Organizations can foster a culture of ethical behaviour by providing training and support for ethical DM, establishing clear codes of conduct and policies, encouraging open dialogue about ethical dilemmas, holding individuals accountable for their decisions and actions, and recognizing and rewarding ethical behaviour.

6.9.3. Environmental, social and governance issues

Sustainability issues are increasingly important to society. Organizations now regularly consider and report on environmental, social and governance (ESG) initiatives to their stakeholders and incorporate ESG considerations into their DM. Societal perspectives of nuclear-related activities have shifted to promote more holistic, inclusive and sustainable DM approaches. This evolution emphasizes the importance of incorporating many diverse aspects and stakeholder views, notably those of civil society [63].

Sustainability encompasses a wide range of issues, including a company's long-term durability as a successful enterprise, climate change and other environmental risks and impacts, financial stability, human capital management, labour standards, resource management, community involvement, and consumer and product safety [64]. Organizations can assist decision makers by including desired ESG considerations within the corporate mission, vision, and value statements, as well as in lower-level procedures.

6.9.4. Transparency and stakeholder input

Stakeholder engagement is crucial for successful DM, with transparency and communication being essential for ensuring stakeholder satisfaction.

Effective stakeholder engagement strategies include implementing consultations, dialogues, and surveys; identifying stakeholders' interests, values, and expectations; recognizing potential areas of collaboration or conflict; learning from the successful practices of other organizations; and clearly communicating important and difficult decisions, especially to affected stakeholders. Regular evaluation and adjustment based on stakeholder feedback are necessary to maintain stakeholder satisfaction. Considering a phased implementation of decisions can minimize short-term negative impacts.

Risk communication is critical for gaining stakeholder support, extending beyond merely conveying information to facilitating dialogue and exchanging perspectives between the organization and its stakeholders. The growing influence of public opinion in nuclear energy development necessitates adopting more transparent and inclusive approaches with the public, providing understandable information that meets civil society expectations.

Local community involvement may be necessary due to regulatory requirements or to ensure the acceptability of decisions. This involvement helps build mutual trust and legitimacy for future projects and decisions. Different types of internal and external stakeholders have varying degrees of influence on DM. These include commercial and financial parties, regulators, supply chains, governments, non-Governmental organizations (NGOs), and the general public.

Understanding stakeholder expectations involves conducting public consultations, dialogues, and networking, as well as performing periodic observations and surveys. Using up-to-date data helps assess the sensitivity and likely satisfaction with decisions. It is important to explain openly and transparently how stakeholder expectations have been balanced in the final decision.

Stakeholder engagement allows organizations to tap into diverse expertise and viewpoints, leading to more comprehensive risk identification, assessment, and mitigation strategies. It ensures broad buy-in and adoption of risk management policies within the organization and facilitates the rapid communication of emerging risks.

Organizations can utilize various resources for stakeholder communication, such as IAEA Safety Standards such as NG-G-5.1 [65], GSR Part 2 [24] and other publications such as Refs [60, 66–68], the Nuclear Communicators Toolbox [6] and OECD/NEA background materials [63]. By implementing these strategies and principles, organizations can effectively engage stakeholders, leading to more informed DM and improved stakeholder satisfaction.

---

[6] See https://www.iaea.org/resources/nuclear-communicators-toolbox

7. DECISION SUPPORT TOOLS AND METHODS

There are a wide variety of tools and methods that support DM. Each depends on having adequate knowledge on which to base the decision (see Section 5.2). The following subsections detail some techniques, methods, and visualization tools available to assist DM.

7.1. DECISION ANALYSIS AND DECISION SUPPORT SYSTEMS

7.1.1. Decision evaluation weighting and decision analysis tools

Alternatives for a decision need to be evaluated logically or rationally to ensure that the correct preferred option is identified unbiasedly. Numerous methods can be employed to account for the preferences assigned to the DM criteria and assign relative weightings. Some methods include expected utility theory[8], multi-attribute utility theory [69], analytical hierarchy process [70] and outranking methods such as those covered in Refs [71, 72].

Decision analysis methods like decision trees, influence diagrams, Bayesian networks, and multi-criteria decision analysis provide prescriptive models for logically mapping out decision scenarios, events, probabilities, and consequences to identify the optimal choice. The details are discussed in Appendix IV.

7.1.2. Decision support systems and software

An increasing set of decision support systems and software tools is available to improve DM quality. Such systems provide data visualization, dashboarding, reporting, and analytics capabilities to support data-driven DM across an organization.

Many general business intelligence platforms like Qlik, Tableau, COGNOS, and Microsoft Power BI can also function as decision support systems. They enable data integration, reporting, visualization, and analytics, providing insights to support organizational DM across various domains.

ERP systems like SAP, Oracle, Microsoft Dynamics, NetSuite, Planview, and many others often have built-in customizable decision support capabilities through dashboards and reporting tools. These dashboards visualize key performance metrics, enable what-if analysis, and provide insights to support decisions around production planning, inventory management, supply chain optimization and other functions.

Logistics and supply chain optimization software analyses demand forecasts, inventory levels, transportation costs, etc., to support procurement, distribution, and inventory stocking level decisions.

Regardless of the platform and whether the tools are specialized for a particular application or general purpose, their core premise is to integrate relevant data, models and analytic capabilities to inform and improve organizational DM processes.

7.2. SCENARIO PLANNING AND WHAT-IF ANALYSIS

7.2.1. Scenario planning

Scenario planning is a method where different future scenarios or situations are envisioned and analysed to understand their implications and impacts on the organization or decision being evaluated. It includes:

- Identifying the drivers of change or uncertainties that could significantly affect the future (e.g., economic, political, technological, and environmental factors)
- Developing narratives or stories around multiple plausible future scenarios based on different combinations of those driving forces playing out;
- Analysing each scenario's risks, opportunities, and threats to test plans/strategies against various circumstances;
- Using the analysis to enhance preparedness and make more robust, resilient decisions that account for different future possibilities and align with strategic plans.

Scenario planning allows organizations to stress-test their strategies, investments, risk mitigation plans, etc., against a range of "what-if" scenarios rather than just expected or baseline projections.

7.2.2. What-if analysis

A what-if analysis is a risk modelling technique that evaluates the impacts on outcomes by changing certain key input variables or assumptions.

For example, in a nuclear project's financial model, a what-if analysis may assess what would happen to profitability if expected electricity revenue declined by 10%, costs increased by 20%, or a 4-month construction schedule delay occurred. This analysis allows an understanding of sensitivities to different risk factors.

Techniques for what-if analysis include Monte Carlo simulations and sensitivity analysis, enabling multiple scenarios to be run by varying input distributions and assumptions to see how outcomes may change.

Compared to scenario planning, which develops narratives around different plausible future states of the world, what-if analysis models the potential effects on outcomes by varying specific variables or assumptions within a model or analysis. Both techniques allow organizations to understand risks better and enhance preparedness through this future-oriented analysis.

7.2.3. Techniques for developing and evaluating scenarios

Table 4 lists some common techniques for developing and evaluating scenarios.

TABLE 4. COMMON TECHNIQUES FOR SCENARIO ANALYSIS

Scenario Analysis Method	Description
SWOT Analysis	SWOT (Strengths, Weaknesses, Opportunities, Threats) analysis can help identify internal and external factors that may shape different scenarios. This analysis provides inputs for constructing scenarios around potential opportunities and threats.
PESTEL Analysis	The PESTEL framework analyses Political, Economic, Social, Technological, Environmental, and Legal factors that could influence future scenarios. This macro-environmental scanning aids in identifying key drivers of change to build scenarios around
Expert Interviews/ Workshops	Engaging subject matter experts, stakeholders, and cross-functional teams through interviews or workshops. The workshop integrates diverse perspectives and knowledge to enrich scenario narratives and assumptions.
Quantitative Modelling	Quantitative techniques like simulations, forecasting models, and sensitivity analysis are used to model different scenarios numerically. These techniques add rigour to scenario evaluation by quantifying potential impacts and testing assumptions.
Scenario Matrices	Plotting the most critical uncertainties on a matrix to derive a set of divergent yet plausible scenarios. This systematic approach ensures scenarios cover a broad range of possibilities.
Pre-mortems and Post-mortems	Conducting pre-mortem analysis by imagining a scenario has failed to identify potential pitfalls. Post-mortem reviews after implementation to capture lessons for improving scenarios.
Back casting	Working backwards from a particular future scenario to determine the events/decisions that could lead to that outcome. This approach tests the internal consistency and plausibility of scenarios.
Storytelling and Visualization	Narrative techniques like storytelling, personas, and visuals make scenarios more vivid and relatable. These techniques aid in effectively communicating scenarios and their implications.

7.2.4. Leveraging scenarios for strategic planning, risk management, and stress testing

Organizations can leverage scenarios for strategic planning, risk management, and stress testing.

In support of strategic planning, organizations can develop multiple scenarios representing plausible future states or operating environments based on key drivers of change and uncertainties (economic, political, technological, etc.). They also can analyse how the organization's existing strategies would perform under each scenario to test their robustness, identify potential vulnerabilities, adjust strategies as needed, and develop contingency plans to enhance preparedness for different futures.

Organizations can integrate scenario analysis into their enterprise risk management framework to enhance risk management. They would construct risk scenarios around potential disruptive events, adverse conditions or "extreme but plausible" situations that could severely impact the organization. Organizations would then:

- Quantify the effects of these risk scenarios on the organization's operations, financial position, supply chains, and other areas through modelling and simulations;
- Identify vulnerabilities, risk exposures and areas requiring mitigation strategies by stress-testing the organization against these scenarios.

Stress testing became an important tool in the nuclear industry in the post-Fukushima period to analyse design cliff edge effects, and it can be used in many applications. Stress testing involves the development of extreme stress scenarios that go beyond expected situations to test the organization's breaking points or failure modes. Organizations can apply financial models, portfolios, business units, designs, or other aspects of these extreme scenarios to evaluate risk bearing capacity. Insights from stress testing can enhance resilience through measures like capital allocation, crisis response planning, risk controls and limits.

Reverse stress testing involves working backward from a business failure scenario to identify causative risk factors and their combinations.

Effective scenario-based strategic planning and stress testing require robust modelling capabilities, strong cross-functional involvement, and an organizational culture that embraces prudent risk management practices. It can provide a competitive edge through enhanced preparedness and adaptability in changing environments when implemented rigorously.

7.2.5. Uncertainty quantification, sensitivity analysis and cliff-edge effects

There may be cases where uncertainty is very high in probability, such as large natural disasters due to a lack of reliable data, dated predictive models, or other reasons. The following methods would help DM in such cases:

- Reducing uncertainties by further research;
- Reducing epistemic uncertainties by collecting knowledge through expert assessment/discussion based on the structured process (such as Senior Seismic Hazard Analysis Committee (SSHAC) process[7] [73] in the case of earthquakes) and by precise modelling;
- Using a "consequence-based (or -biased) approach" instead of focusing on probability;
- Evaluating "where is cliff edge" by stress test or calculating conditional probability in PSA (See Section 3.4.3 of Ref. [2]).

The risk management principles of INPO [20] describe a "consequence-biased approach" wherein unacceptable consequences are considered intolerable in safety-related DM, even if their probability is low.

At an NPP, a PSA will assess an event scenario, its probability and consequence, and will be the basis for various safety-related decisions. Learnings from the Fukushima Daiichi accident that have been incorporated into IAEA Specific Safety Requirements for Design SSR-2/1 Rev 1 [27] have established the need to consider abrupt changes in potential consequences and the identification of cliff-edge effects.

Decision making approaches to determine actions (including owner/operator self-driven actions) in the aftermath of the Fukushima-Daiichi Accident are documented in Section 3.4. of Ref. [74], including value impact assessment.

7.3. SIMULATION AND COMPUTATIONAL MODELLING

Simulations and computational models can provide valuable information for decisions related to nuclear facilities. Full scope and desktop simulators and digital twin systems can test potential design, operational or maintenance changes. Engineering models of various structures, systems and components (SSCs) can look for unanticipated impacts of changes, analyse

---

[7] SSHAC is an expert panel organized to discuss epistemic uncertainties in the inputs to hazard analysis for PSHA (Probabilistic Seismic Hazard Analysis)

accident scenarios, analyse thermal-hydraulic impacts, model ageing effects, model fuel burnup, and provide for high-quality designs and engineering decisions.

Models can also be developed to analyse business processes, transportation and logistics, supply chains, project schedules, risk and scenario impacts, and organizational structures. Approaches can include system modelling approaches such as discrete event modelling, agent-based simulation, system dynamics, and queueing models, as well as computational techniques such as Monte Carlo methods, pseudorandom number generators, and others.

Simulations and models need to be validated and verified before they are used in critical DM. Once developed, their use needs to be integrated into organizational DM processes where they apply.

7.4. PARTICIPATORY DECISION MAKING TECHNIQUES

Silent DM meetings miss valuable dialogue opportunities between people with different views and solutions, through which better decisions may be attained. Participatory DM techniques ensure that concerned individuals or groups are involved in the DM process. These techniques range from minimal participation to full participation and can be categorized into several types:

- Consultative decision making: In this approach, the DM authority resides with the manager, who may consult or seek opinions from other stakeholders. This method allows for some input but maintains the manager's final authority.
- Democratic decision making: In this approach, the decision is made by the group as a whole rather than by an individual. This method ensures that all stakeholders have an equal say in DM.
- Delphi technique: This method involves a panel of independent experts who provide anonymous forecasts over multiple rounds. The process stops when the forecasts show minimal variation between rounds, and the final round forecasts are averaged. This complex and time-consuming technique provides more accurate forecasts than unstructured groups.
- Brainstorming: This technique involves people generating ideas and solutions to a problem. It is often used to create a wide range of ideas and can encourage creativity and participation. Smaller groups can be used to encourage participation.
- Voting and ranking: These methods involve participants voting or ranking options to determine the best course of action. They can be used to ensure that all stakeholders have a say in the DM process.
- Consensus building: This method involves finding a solution acceptable to all stakeholders. It requires active listening, open communication, and a willingness to compromise.
- Participative decision making: This approach involves all stakeholders in the DM process, ensuring everyone has a say. It can be time-consuming but can lead to more engaged and committed participants. The ODM process described in Section 5.1.3.3 is an example of this technique.

Participatory processes can suffer from groupthink if not carefully managed. Methods to avoid groupthink include the brainstorming technique described above, as well as:

- Challenging the status quo: Start with questions about the established norms. Invite external advisors for different perspectives. Create a safe space and encourage positive feedback with any suggestions.
- Critical review: Encourage every participant to review the proposed options critically. Use anonymous evaluations to minimize bias.
- Use of small groups: Dividing a meeting into small groups to solicit diverse views,
- Devil's advocate: Assign a "devil's advocate" [75] to pose critical questions or provide an alternative view. A decision sheet records questions raised by "devil's advocate" and related discussions before concluding.
- Focus broadly: Focus on broad principles and goals, allowing for diverse perspectives and approaches and a shared vision.

Additionally, stakeholder engagement processes (See Section 6.9.4) assist in bringing different views and perspectives into DM and, thus, are a form of participatory DM.

7.5. ADDRESSING TIME PRESSURE

Time pressures increase the potential for decision errors because rational thinking or collecting necessary information or deliberation is hampered by limited time and stress. Consequently, decisions based on intuition and past experiences (especially those of the senior managers) may dominate.

The type of decision needed impacts how time pressure might affect a decision. As discussed in Sections 3.4 and 3.5, there are many DM styles and relevant organizational structures. Some approaches are systematic, involving gathering information, considering alternatives, communicating with stakeholders, and carefully considering decisions. Such approaches take time. Other methods are instinctual—involving heuristics and prior experience for quick decisions. The challenge is that some situations require quick DM due to time pressure, but heuristics may not serve them well. Development of scenario-based, 'safe-to-fail' exercises that build awareness and consequent DM capability are strongly recommended. There can often be a speed-accuracy trade-off in DM, yet building skills through training is one way to off-set these impacts.

Sometimes, decision makers feel pressured by time, even when the situation allows for a more thorough analysis. Conversely, decisions are sometimes delayed even when a faster response is needed. Defining a clear timeline is one of the crucial first steps in the DM process. It is important to determine whether a quick decision is necessary or if a longer, more considered decision is preferable.

7.5.1. Thinking fast and slow

Nobel Economic Laureate Daniel Kahneman summarized research on DM into the concept of Thinking Fast and Slow [91]. System 1 thinking involves fast, automatic, and intuitive decision processes. System 2 thinking involves slow, conscious, and effortful decision processes.

The two types of DM are complementary and have strengths and weaknesses, as depicted in Table 5. A misconception is that thinking fast pertains only to operational activities, while thinking slow applies to strategic matters. Higher consequence DM has to rely on systematic processes embodied by "thinking slow." Still, senior managers may be called on to make tactical or operational decisions quickly and require "thinking fast."

TABLE 5. ADVANTAGES AND DISADVANTAGES OF THINKING "FAST" VS. "SLOW"

Thinking System	Advantages	Disadvantages
System 1 "Thinking Fast"	Organic Answer—The general DM process holds true much of the time without costly effort	Biases—The use of shortcuts can lead to suboptimal conclusions.
System 2 "Thinking Slow"	Systematic Result—Consideration of multiple points of view with trade-offs to arrive at the best or consensus decision.	Decision Paralysis—The process involving weighing may not meet urgent time needs for certain decisions

The challenge is that under time pressure, such as when a decision needs to be made immediately, we might feel there is not enough time for 'Thinking Slow.' It is an evolutionarily developed human behaviour to rely on 'Thinking Fast' when under pressure. However, this may introduce pitfalls such as biases to do something that worked before. That previously used solution may not always be the best solution for the current problem, and alternatives that would more effectively solve the problem may be overlooked. Thinking fast may thus not always produce the most effective decisions. Especially when significant risks are involved, decisions need to be well-founded, allowing time for a brief evaluation. Operating procedures or checklists typically cover actions that need to be taken quickly. In any case, the decision maker has to deliberately engage System 2 thinking under time pressure to make rational decisions.

System 1 and 2 thinking mostly applies to individual DM. An effective approach to avoid the pitfalls of biases is to make use of the organization to support the DM. The organization has to be able to enlist expertise and knowledge quickly to support time-sensitive decisions. A few key steps include:

— Identify the main factors at play, such as safety, plant systems, and environmental impacts that may be at stake. Prioritize these and develop clear objectives for the decisions. Decision makers may not always be aware of the time pressures or other limitations associated with a decision. Guidance in procedures or checklists (see Section 5.1.4) regarding the necessary timelines for key decisions can help to reduce such inappropriate choices;
— Establish communication channels, deadlines and milestones to ensure information is communicated effectively and the DM process is not delayed;
— Identify and enlist key stakeholders to provide key and timely decisions. The stakeholders are the main information holders who can accelerate DM. This group is a limited set of individuals who can speed up DM rather than bog it down;
— Avoid unnecessary meetings that will distract from the DM;
— Don't get lost in the forest of decisions; focus on the most high-impact ones. Smaller, low-impact decisions have to be made quickly;
— Continuously monitor the outcomes of the decisions. Assess those milestones and determine if decision pivots are necessary, such as when a solution is not working and alternatives need to be considered. Poor decisions have to have an opportunity for correction;
— Use proven DM frameworks such as strengths-weaknesses-opportunities-threats (SWOT) or cost-benefit analysis;

– Provide individuals with appropriate training or drills in DM before the need (see Section 5.3.3).

7.5.2. Hybrid approaches

Decision making is rarely perfect the first time it is tried. It will be helpful to practice DM under time pressure through drills and training exercises. Such drills can also speed the process and ensure there is adequate time not to make hasty or impulsive decisions but rather to consider the information and possible outcomes efficiently. This approach would be a hybrid of System 1 and 2 thinking. Many methods can address time pressures that combine fast and slow thinking approaches. Table 6 below provides some examples.

TABLE 6. HYBRID APPROACHES TO ADDRESS TIME PRESSURE

Method	Description
Planning and Rehearsing	Anticipating problems in advance can forego the need for time-consuming, on-the-fly information gathering. Rehearsing and simulating emergent issues can prepare the decision maker for such situations if they happen.
Accelerated Walkthroughs	Gathering experts together quickly for an accelerated walkthrough of the steps in Section 3.1 can gather crucial information and viewpoints for DM in a single session. This approach ensures decisions are made systematically, quickly and with all information at hand.
Leveraging PSA results	Where competing options are present, using risk informed outcomes from the PSA (if available) can help determine the safest, least impactful action.
Decision Mileposts	Breaking the decision into "mileposts" where certain actions can immediately proceed while allowing time for more systematic System 2 DM of other actions. Mileposts can provide proof points to reevaluate the outcomes of decisions in progress and pivot the decision if needed.
Build in time for decisions	Build sufficient margin and flexibility in business processes to ensure resilience in the face of DM requirements.
Leverage technology	Leverage emerging technologies (e.g., predictive maintenance systems, prognostic operator support systems, or generative AI) to provide additional information to help with diagnosis and DM. In many implementations, these tools can identify problems and recommend solutions that benefit the decision maker and may allow more rapid and informed DM than has been previously possible.

7.6. VALUE-IMPACT ANALYSIS

Very often, decision makers wonder if the cost of a proposed modification is commensurate with its benefits. A generic way of expressing "cost vs. benefit" is "value vs. impact."[8]. Neither value nor impact limits itself to monetary value. Intangible benefits such as reducing operator burden or improving corporate reputation also need to be considered.

Value-impact analysis compares the monetary cost (the "impact") of the proposed action and its benefits ("the value"). Value is defined as the beneficial aspects anticipated from a proposed action, such as, but not limited to:

---

[8] Stopping operation of isotope producing reactor NRU due to concern over safety has "value" for residents' safety but has "impact" of reducing medical isotope supply globally.

- Protection of public health and enhancement of public safety;
- Protection of the natural environment;
- Promotion of the efficient functioning of the economy and private markets;
- Elimination or reduction of discrimination or bias;
- Enhancement of plant or operating teams' performance and reliability.

Impact refers to the costs anticipated from a proposed action such as, but not limited to:

- Direct costs to affected organizations in implementing or complying with the proposed action;
- Adverse effects on public health and safety, and the natural environment;
- Adverse effects on the efficient functioning of the economy or private markets;
- Increased complexity of plant operation or maintenance.

Some examples where value-impact analysis principles apply or have been invoked within the nuclear industry are in Table 7. Appendix VI provides more detail on some of the examples.

TABLE 7. VALUE IMPACT ANALYSIS EXAMPLES

Topic	Source and Commentary	Reference
Radiation Protection	ALARA Principle described in ICRP Publications 9 (para. 52) and 22 (para. 11)	[76, 77] Appendix VI.1
Assessment of Health Impacts	Mandate to consider the impact on the health of Canadians who, for medical purposes, depend on nuclear substances produced by nuclear reactors. Issued to address the unplanned shutdown of the NRU Research Reactor that supplied a large percentage of the global Mo-99 medical isotope.	[78]
Cost Benefit Analysis Related to Backfitting	The US NRC weighs the costs of implementing backfitting vs. the safety and security benefits of the proposed change. NUREG/BR-0184 (Regulatory Analysis Technical Evaluation Handbook) [79] and NUREG-1530 [80] provide guidance for valuing dose savings ($2,000/person-rem (See Ref. [18])). NUREG/BR-0184 [79] and NUREG/CR-3568 [81] include information on calculating the value of property damage.	[82] Appendix VI.2

TABLE 7. VALUE IMPACT ANALYSIS EXAMPLES

Topic	Source and Commentary	Reference
EDF (France) assessments of potential modifications relate to the Long-Term Operation (LTO) of the French NPP fleet	The DM process is implemented by considering all relevant factors (safety of the design, human and operational factors, security, environment protection, and costs) to select and prioritize appropriate design and operational options when improving plant safety levels. A Safety Issues Assessment method was developed with other nuclear industry operators, which relied on a set of criteria and items reviewed by a multi-skilled, multi-disciplinary group.	See the Appendix V in Ref. [2]

7.7. ARTIFICIAL INTELLIGENCE IN DECISION MAKING

Artificial Intelligence (AI) applications are being explored and tested in the nuclear industry to assist decision making. Applications include predictive maintenance and condition monitoring systems, knowledge management and retrieval systems, design optimization, radiation monitoring, digital twins, advanced control room operations, cybersecurity, fuel optimization, supply chain optimization, robotic inspections and others. Such AI systems need to be transparent, explainable and certifiable to gain the trust of operating organizations and regulators. General industry is beginning to adopt AI in making strategic decisions [83].

AI can help DM in many ways:

- Increasing speed (AI systems can process data and generate insights very rapidly, sometimes in real time);
- Increasing productivity and efficiency (ability to work 24/7 on difficult issues, ability to identify process inefficiencies);
- Improved accuracy (AIs can identify complex patterns, correlations, and anomalies that humans might overlook);
- Reducing risk (AI algorithms can detect patterns and anomalies that indicate potential risks, such as fraudulent activities, market fluctuations, or supply chain disruptions);
- Improved consistency (AIs can apply predefined rules to ensure decisions are made consistently);
- Improved Knowledge Management (AIs can maintain a repository of organizational knowledge).

The Canadian Nuclear Safety Commission (CNSC), the United Kingdom's Office for Nuclear Regulation (UK ONR), and the United States Nuclear Regulatory Commission (US NRC) have issued a collaborative publication that outlines high-level principles for the deployment of artificial intelligence in nuclear applications [84]. The document focuses on operational as opposed to strategic decision making.

AI systems can introduce new risks, such as increased reliance on complex algorithms and the potential for unforeseen errors. It is essential to manage these risks proportionately, ensuring that AI deployment does not unduly stress reactor safety, security or other high-consequence areas. For example, an AI system for analysing maintenance data needs to be carefully evaluated to avoid unintended consequences like increased spurious equipment trips.

Figure 3. provides a method for considering the overall AI system risk [84]. The most risk is present with those AI systems with the most autonomy (least human interaction) coupled with the highest consequences of failure. A similar approach could be applied to consider AI systems that could impact other high-consequence areas (e.g. strategic, business or financial AI models).

The regions within the figure with high failure consequences (regions 3 and 4) are the regions of most concern. AI systems deployed within region 4, the region of high autonomy and high consequence, are also characterized by a low ability to verify a system's output before action.

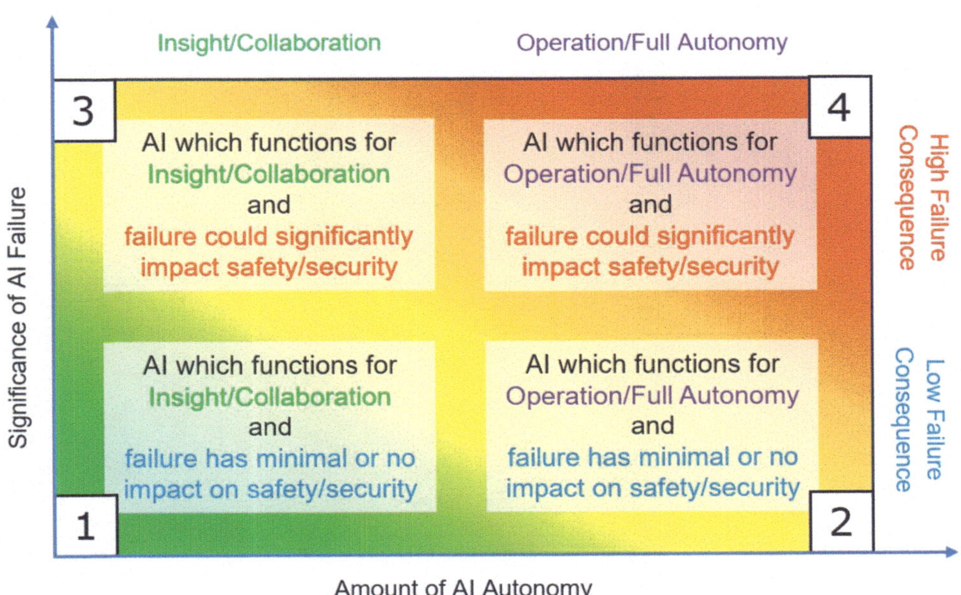

FIG. 3. Categorizing AI failure significance and AI autonomy [84]

Some key methods to minimize risk when deploying AI in nuclear power applications include:

— Prioritize simplicity: AI may overcomplicate simple syatems with little added benefit, while combining AI with simpler technologies may help prevent unknown failures;
— Apply established engineering principles for managing nuclear risk (e.g., diversity, redundancy, separation, segregation);
— Ensure proper human factors integration (operator training, managing trust in the outputs of AI systems while guarding against complacency, and ensuring model biases in DM are addressed);
— Maintain thorough documentation (of AI development, testing, changes, and verification);
— Carefully design AI architecture (model selection, data quality, testing);
— Implement robust lifecycle management (monitoring, updating, maintaining the AI, and continuously evaluating and retraining the AI with new data as needed).

Appendix VII of this publication provides some examples of AI use in DM in France (a digital twin of the NPP with AI), Russia (a digital assistant within its corporate decision making centre), and the USA (various AI applications, including smart searches, PSA improvements, troubleshooting and training).

8. CONCLUSIONS

Nuclear power plant operating organizations routinely face challenging strategic and tactical decisions. Correct DM is often challenged by factors such as insufficient information, large uncertainties, time pressure, business conditions, issue complexity, organizational culture/traits, and human elements such as heuristics and biases. Enhancing the structured DM process across all organizational levels is crucial for their success and resilience.

Key success factors include:

- Ensuring everyone understands the fundamentals of decision making (DM), including the structured DM process, potential pitfalls, methods, skills, and tools. This understanding can be achieved through a training program integrated into the management system;
- Clearly defining the DM process, decision accountability and authority, and the roles of DA and independent oversight within the management system;
- Utilizing and referencing risk informed DM methods and tools already in use and documented at numerous nuclear power plants (NPPs);
- Strengthening managers' DM skills, such as communication, sense-making (capturing and sharing the status quo with the team), and emotional intelligence for leading team members. Additionally, providing guiding principles rooted in nuclear safety, security, and the organization's vision, mission, and values;
- Recognizing that organizational culture and traits can influence decision outcomes, either positively or negatively, and analysing past flawed decisions to enhance decision quality;
- Developing and maintaining a knowledge management process to support informed DM;
- Leveraging AI to assist DM while ensuring that final decisions remain under human judgment;
- Using diverse methods/tools to support DM, including probabilistic safety assessment (PSA), stress testing, what-if analysis, and scenario analysis;
- Addressing high-uncertainty issues, particularly those involving natural hazards with low probability, by conducting further research, gathering knowledge, organizing expert discussions, creating precise models, adopting a consequence-based approach over a probability focus, and identifying "cliff edges" through stress tests or conditional probability calculations in PSA to reduce uncertainties effectively;
- Mitigating the likelihood of decision errors during time-sensitive situations by acknowledging the impact of limited time and stress on rational thinking, information collection, and deliberation. DM by a team can help identify and counteract biases more effectively.

APPENDIX I.
PSYCHOLOGICAL TRAPS, HEURISTICS AND BIASES

Even the smartest and most experienced person makes mistakes. Consequently, decision makers have to be trained and aware of how people make mistakes and how to avoid them.

The pitfalls in DM often arise from psychological traps, heuristics, and biases. Individual organizations may have unique traits (See Section 5.3.2), which decision makers are unaware of but can be recognized by analysis of past mistakes in DM.

When uncertainty is high or where there is insufficient information, decision makers tend to make biased judgements due to mental shortcuts to reach a quick decision (*heuristic*s) [14]. Heuristics is a term used in psychology to express a way of intuitively obtaining a quick answer through experiences accumulated in an individual or a group and preconceptions. Heuristic-driven cognitive biases lead to mistakes in decisions. Table 8, adapted from Ref. [15], lists typical heuristics and biases divided into five bias families, plus selected examples from the nuclear industry.

TABLE 8. HEURISTICS AND BIASES [15]

Bias Family	Bias	Definition
Pattern Recognition	Confirmation; storytelling; framing effects	Paying attention to facts that support one's hypothesis, particularly when well-organized. Different preferences are based on how choices are framed.
	Experience / Availability	Reasoning by one's own part experience. Judging probability by how easily instances come to mind or how retrievable they are.
	Champion	Giving too much weight to the reputation of the messenger.
	Attribution error	Attribute success to individual action as opposed to chance
	Hindsight	Judge past decisions based on information that was not knowable at the time,
	Halo Effect	Form impressions based on a few features and let that impression influence the assessment of unrelated features.
	Survivorship	Base decisions based on successes but not failures.
Action-oriented	Over placement	Overestimating relative ability as compared to others.
	Planning fallacy; unrealistic optimism	Not considering items or combinations of items that could derail plans, resulting in optimistic estimates of time or cost to complete.
	Over precision	Overestimating degree of confidence in forecasts.
	Competitor Neglect	Not considering the actions of competitors in a decision,
Inertia	Anchoring; Representativeness	Relying too heavily on an initial piece of information or by available numbers, even if they are irrelevant. Judging probability is based on stereotypes or limited data that might not be representative, valid, or have a low sample size.
	Resource inertia	Being timid in reallocating resources to address priorities.
	Status Quo	Avoiding a decision or change and maintaining the status quo/
	Escalation of commitment/ sunk cost fallacy	Sticking with failing courses of action, not treating previous investments as sunk costs.
	Loss aversion; Hyperbolic Discounting	Feeling losses more keenly than gains. Humans exhibit a preference reversal regarding timing choices, discounting smaller sooner rewards much more than larger later rewards, contradicting the exponential discounting predicted by normative models.

TABLE 8. HEURISTICS AND BIASES [15]

Bias Family	Bias	Definition
Social	Groupthink	Groups silencing doubts and individuals siding with prevailing opinions instead of dissenting,
	Polarization	Groups reach conclusions that are more extreme than their average view.
	Information cascades	The sequence of speakers affects the outcome of discussions.
Interest	Self-serving	Prefer viewpoints that align with one's own interests.
	Present	Using inconsistent discount rates, overweighting presents short-term benefits.
	Omission	More indulgent of omissions of others than of errors of commission.

Biases can have many negative impacts, including flawed or distorted risk assessments, negative perceptions of nuclear power among the public and policymakers, and poor decisions.

Awareness of these cognitive biases and applying debiasing strategies is crucial for improving risk literacy, public trust, and DM around nuclear energy and safety.

APPENDIX II.
THE FUKUSHIMA DAIICHI ACCIDENT IN THE CONTEXT OF DECISION MAKING

Most authoritative reports on the accident do not explicitly discuss an analysis of decision-making nor the heuristics and biases behind it. Nevertheless, it would be worth including short discussions on this topic so readers may learn something from it.

II.1.   BEFORE THE ACCIDENT

II.1.1.   Consequence-biased approach in decision making

In hindsight, it is clear that a consequence-biased approach (Ref. WANO GL-2002-01 Rev.1 [3]) was not taken to protect against a nuclear accident caused by a natural disaster. In 2008, a tsunami inundation height of 15.7 meters was calculated based on a hypothetical offshore source near Fukushima. A subsequent stress test, which incrementally increased the predicted tsunami height, indicated severe damage to the plant's safety systems, including loss of power and Emergency Core Cooling System (ECCS) actuation. However, no decision was made in response to these findings—perhaps due to the perceived low probability of such an event, high uncertainty, and the influence of authoritative seismologists' groupthink.

II.1.2.   Groupthink trap

Although the government's Headquarters for Earthquake Research Promotion (HERP) reported in 2002 that an earthquake might hit anywhere along the northeast coast of the Pacific Ocean, the government's Disaster Management Plan was not amended, nor was any regulatory action taken. HERP's website indicated immediately after the earthquake that the occurrence of the earthquake within all of these regions is "out of hypothesis."

In a paper for the International Symposium (March 1-4, 2012) on Engineering Lessons Learned from the 2011 Great East Japan Earthquake, Professor Matsuzawa of Tohoku University wrote, "There were many evidences showing weak inter-plate coupling in the north-east Japan subduction zone: age of the descending slab, background seismicity, existence of small repeating earthquakes, long-term strain in the land area, etc. Thus, almost all seismologists did not anticipate that M9 earthquakes could occur there. [85]"

Although there seems to have been limited criticism that this local subduction theory was supported by only 100 years of observation data and that the 2004 Sumatra earthquake and tsunami did not support this theory, TEPCO's decision maker may have been influenced by this groupthink trap.

II.1.3.   Possible status quo bias and loss aversion bias in decision making

Given that there was increasing concern over a tsunami along the coastline facing the Pacific Ocean, TEPCO's seven NPPs (8.2GWe) on the Japan sea side were lost by the 2007 earthquake, stopping the operation of six units at Fukushima-Daiichi (4.7GWe) to address the tsunami concern to install large scale modifications (such as moving underground electric equipment room or Emergency DG to higher location) may have intimidated decision makers due to a "loss aversion bias" leading to a "status quo bias." Also, the high uncertainty in the probability of the threat's happening may have led to a delayed decision.

II.1.4.   PSA of external events to support decision making

Japan did not use a PSA to support decision making regarding external hazards. Although the country had a history of many natural hazards (earthquakes, tsunamis, typhoons), the general attitude against an external event PSA by operators was that "it is not useful due to high uncertainties." A tsunami PSA only became available within TEPCO in 2006. Even so, the response by TEPCO was poor, as was lamented by its ex-CNO later [86]. Unlike internal events, it is natural that uncertainty is high for external events. However, this high uncertainty is a fact and is not because of the immaturity of PSA techniques. The decision maker did not take those actions described in section 7.2.5 of this publication.

II.1.5.   Possible overconfidence bias in decision making

Japan had seen a low frequency of scram and high reliability of components by intensive preventative maintenance, which possibly had led to operator's overconfidence in safety since "high-reliability component assures safety." SAM (severe accident measures) was taken in the 1990s for level 4 Defence in Depth such as hardened venting of containment, enabling injection from low-pressure makeup system (such as fire protection system) to the Reactor Pressure Vessel (RPV) to supplement ECCS, installation of air-cooled Emergency Deisel Generator (EDG) for diversity etc., which led to the thought that safety has been assured (although SAM measures were postponed to counter external events). This overconfidence may also have been the reason for thinking "accident will not happen here" and for not learning from past flooding of the underground floor of the turbine building of Fukushima Unit 1 due to seawater leakage (1991) or Blayais flooding (France, 1999) or Kalpakkam NPP site (India, tsunami by Sumatra earthquake, 2004).

II.1.6.   Competence and capability for informed decision making

Informed decision making requires knowledge of and expertise in NPP operation and management of decision making. Some reports (e.g., [86]) point out the lack of knowledge and expertise by TEPCO and regulators (including academia supporting it) on such issues as tsunami, PSA, Isolation Condenser design, and risk communication. One example is that the US B5b regulation [9] to cope with severe conditions at NPPs was conveyed to Japanese regulators confidentially. However, the receiving regulators in Japan could not understand it and did not take the necessary actions.

II.1.7.   Autonomy vs. corporate governance of nuclear division on decision making

After the accident, WANO modified its peer review service to address the issue of weak corporate governance over nuclear organizations and its own governance related to WANO regional centres.

---

[9] The US B5b regulation, established by the US NRC, was introduced in response to the events of September 11. 2001. It requires NPPs to develop and implement strategies to maintain or restore core cooling, containment, and spent fuel pool cooling capabilities in the event of a loss of large areas of the plant due to explosions or fire.

II.2. DURING THE ACCIDENT

II.2.1. Confusion in decision authority

During the accident, there were cases of confusion in decision making authority. Examples are the intervention of the Prime Minister on containment venting issues (later, Japanese law was amended to prevent such interventions) and the government's Cabinet Office [86] and the confusion of authority within TEPCO between its headquarters and the plant manager in deciding if water injection to the RPV needs to be stopped or not.

II.2.2. Groupthink trap

According to an INPO report [87], the decision making approach applied during the accident did not provide for independent challenges or second checks by other groups within the organization, as was seen at the site emergency response centre.

II.3. BUSINESS BACKGROUND

Figure 4. below shows the business background that existed in TEPCO then and influenced DM. It displays a negative spiral that results in the degradation of three fundamental capabilities required in NPP operation and management – safety awareness, engineering and technical capabilities, and communication capability.

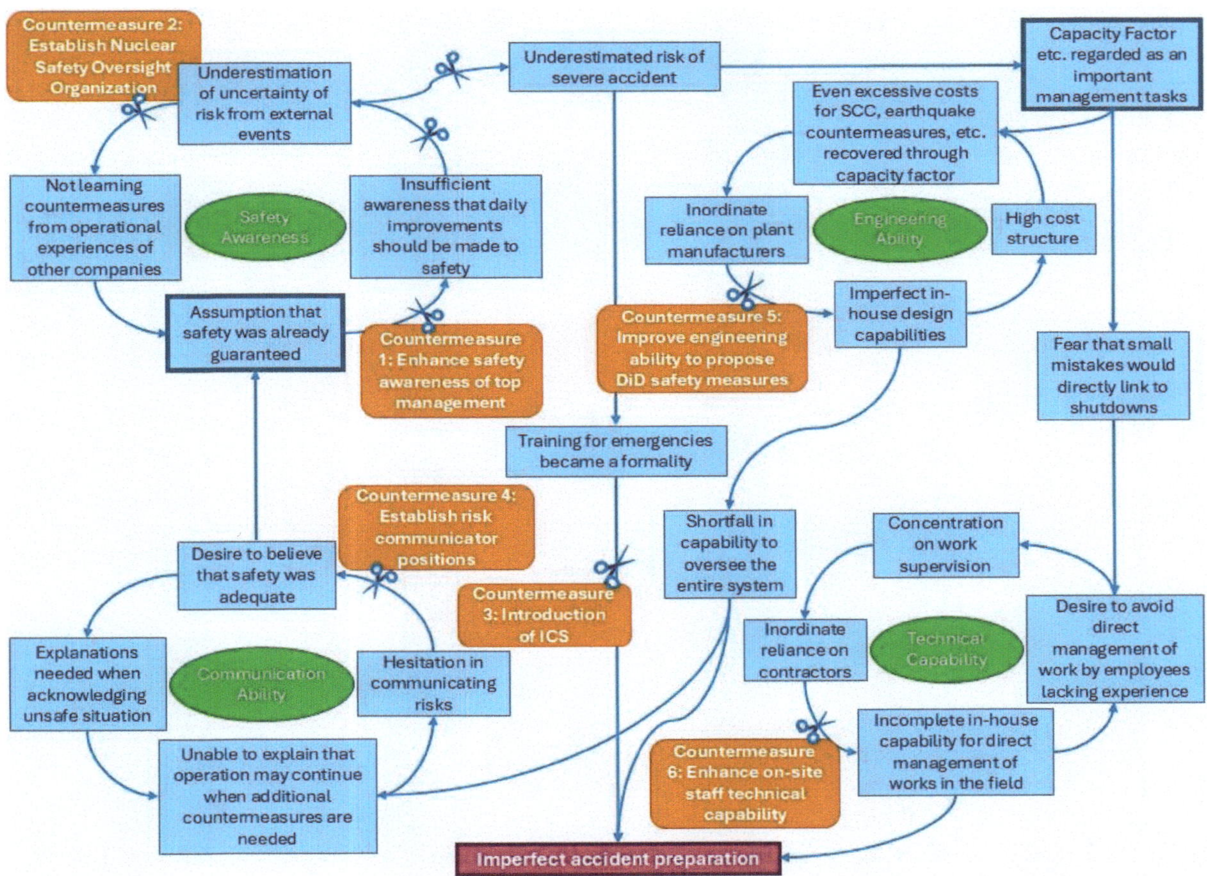

FIG. 4. Business background – Negative spiral that caused the accident

The spiral starts from the belief and assumption that safety has already been established. It was gradually embedded in TEPCO's organization after the SAM implementation in the 1990's. TEPCO also had an environment where top management prioritized improving NPP capacity factors. The TEPCO scandal in 2002 and the Nigata-Chuetsuoki earthquake in 2007 both had caused long periods of plant shutdown. While the vision was safety first, safety awareness started to degrade because of this focus on capacity factor.

The focus on improving capacity factors increased the outsourcing of design and modification work to manufacturers like Toshiba and Hitachi and field work to contractors since TEPCO feared that its small mistakes could cause plant shutdowns. The focus resulted in engineering and technical capability degradation and created a high-cost structure. It further affected its communication capability with public and local government stakeholders. TEPCO repeatedly stated, "Nuclear is safe and perfect. Accident will never happen." without any detailed technical discussion. Further details are discussed in Ref. [88].

APPENDIX III.
RISK ASSESSMENT METHOD WITHIN NUCLEAR

This Appendix describes common methods to evaluate various risks within a nuclear environment. Such evaluations are essential in making decisions related to nuclear siting, design, and operation and maintenance activities. Strategic decision makers need to know these techniques as part of the detailed steps that contribute to the final global decision.

III.1. PROBABILISTIC SAFETY ASSESSMENT METHODS FOR NUCLEAR FACILITIES

The advent of PSA technologies, coupled with the accumulation of data on initiating events and component reliability, has made using risk information in DM increasingly popular. Recent improvements in PSAs have been made post-Fukushima to better account for external hazards, human reliability in harsh environments, the behaviour of containment and radioactive nuclides, and multi-unit impacts.

The IAEA has published numerous safety standards and other publications related to the development and application of PSAs, including SSG-3 [89], SSG-4 [90], SRS-25 [91], TECDOC-1101 [92], TECDOC-1200 [93], TECDOC-1267 [94], TECDOC-1511 [95], and TECDOC-1977 [96]. Other standards organizations have also published guidance, such as ASME/ANS's RA-S-1.1 [97], RA-S-1.2 [98], RAS-1.3 [99] and RA-S-1.4 [100], and CSA N290.17 [101].

Using risk information (especially a plant-specific PSA) as an integral part of DM is highly recommended for issues important to safety. Consideration has to be given to the following areas when using such risk information:

- Quality and scope of PSA;
- Other considerations and analyses, such as deterministic assessments, regulatory requirements, cost-benefit analysis, engineering judgement, and subject-matter expert opinions, could inform the DM process;
- Principles and attributes for excellence in risk informed risk management as are discussed in INPO 15-011 [20] Principles for Excellence in Integrated Risk Management, such as consequences that need to be assessed from the perspective of external stakeholders;
- Assumptions in PSA, such as the exclusion of low probability initiating events from the analysis;
- Uncertainties;
- Safety goals;
- "Practically eliminate" (new build) or radically reduce early and large releases and related cliff edge effects (discussed in IAEA Specific Safety Requirement (SSR) -2/1, Rev.1 [27] Section 2.13 and 5.73).

III.2. FAILURE MODES AND EFFECTS ANALYSIS

Failure mode and effect analysis is a systematic risk analysis technique used to identify potential failure modes in a system, product or process, evaluate their causes and effects, and prioritize areas for corrective actions. It is a qualitative or semi-quantitative analysis method complementing quantitative risk assessment approaches like PSA.

Failure mode and effect analysis provides a list of potential failure modes and analyses the possible causes and effects on each system. Risk priority numbers (RPNs) are typically calculated based on ratings for the severity of effects, the likelihood of occurrence, and the likelihood of detection. Higher RPNs identify the most critical failure modes to prioritize for mitigation actions.

Failure mode and effect analysis can be applied at the system level (System FMEA), product/design level (Design FMEA) or process level (Process FMEA).

III.3.   HAZARD AND OPERABILITY ANALYSIS (HAZOP)

Hazard and operability analysis identifies hazards to personnel, equipment, and the environment and operability issues that impact operational efficiency. It involves breaking down the system into nodes for individual review by a multi-disciplinary team. The team then identifies the causes and consequences of potential deviations and identifies actions necessary to mitigate identified hazards or operability issues. Some reference publications include IEC 61882 [102] and API RP14C [103].

Hazard and operability analysis is most often applied during a design phase. Other methods, such as constructability, operability, maintainability and safety (COMS) reviews address similar concerns.

III.4.   HUMAN RELIABILITY ANALYSIS (HRA) METHODS

Human reliability analysis (HRA) involves using qualitative and quantitative methods to assess the human contribution to risk. Human reliability analysis considers different types of human actions - pre-initiators (errors during maintenance/testing), initiators (errors causing an initiating event), and post-initiators (errors in responding to events). Various methods have been employed in industry, many supporting Level 1 or 2 PSAs. Common methods include the technique for human error rate prediction (THERP), standardized plant analysis risk-human (SPAR-H), Korean HRA (K-HRA) and cognitive reliability and error analysis method (CREAM). Ref. [104] provides a review of some of these methods.

III.5.   EXTERNAL HAZARDS ANALYSIS (EHA)

External hazard analysis (EHA) refers to the process of identifying, evaluating, and mitigating potential risks posed by natural or human-induced hazards originating from outside a facility or site. It covers many external hazards such as earthquakes, floods, hurricanes, tsunamis, tornadoes, volcanic eruptions, wildfires, aircraft or railroad crashes, toxic releases from nearby facilities, and other severe weather events. PSA techniques quantify the likelihood of these hazards and the potential consequences on the facility's safety systems, structures, and components.

The IAEA provides numerous safety standards and other publications related to EHA, including SSG-77 [105], SSG-68 [106], NS-G-2.13 [107], Safety Report No. 92 [108] and 94 [109]

APPENDIX IV.
DECISION ANALYSIS TECHNIQUES

IV.1.   DECISION TREES

Decision trees are graphical representations of DM processes and can assist in visualizing the considerations that lead to each possible outcome. They break down decisions into a sequence of choices or events. Each node in the tree represents a decision point or an event. Each branch represents a possible outcome or choice. Figure 5. provides a simple example.

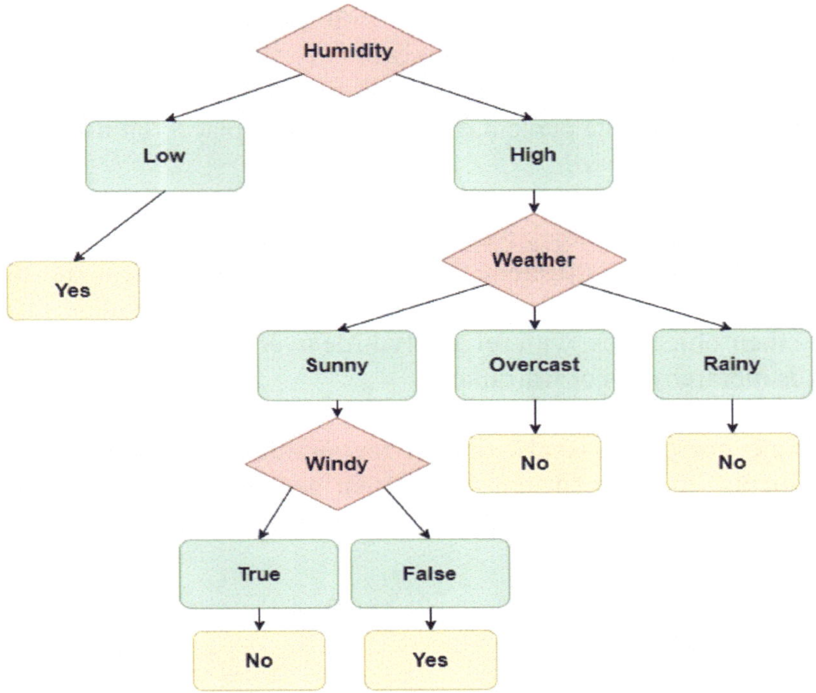

FIG. 5. Simple Decision Tree to decide whether to play outside

IV.2.   INFLUENCE DIAGRAMS

Influence diagrams extend decision trees by incorporating additional elements like uncertainties, objectives, and dependencies. In the example in Figure 5., each node could be given a probability of its occurring (e.g. 80% chance of it being sunny), and an overall probability of a decision to go outside or not could be determined.

IV.3.   BAYESIAN NETWORKS

Bayesian networks (BNs) represent probabilistic dependencies among variables. Nodes in the graph represent variables, and edges indicate probabilistic relationships.

IV.4.   MULTI-CRITERIA DECISION ANALYSIS (MCDA)

Decisions are often made by groups of individuals working together. These individuals may have different objectives or differences in opinion concerning what the best decision would be and what trade-offs they may prefer. It becomes important to structure a DM process to account for these differences and to make the necessary trade-offs needed to come to a decision.

Constructing a decision hierarchy can be useful in these circumstances. The steps to develop such a hierarchy in a group setting can include:

- Brainstorm different factors (or evaluation concerns) involved in the decision;
- Specify the direction (maximize/minimize/maintain) that improves attaining the objective;
- Construct an objectives hierarchy by placing the objectives generated into a tree structure;
- List the consequences of alternatives against each objective. A useful way to illustrate the decision is in a spreadsheet model, with the rows being the objectives and the columns corresponding to the alternatives;
- Assign to each lowest level objective a weight that indicates how important it is (each swing weight defining the percentage of importance placed on that objective).

When multiple stakeholders come to a decision, sometimes a single objectives hierarchy can be structured. In other cases, an objectives hierarchy will be constructed for each stakeholder because their objectives are so different that the construction of separate hierarchies more accurately represents their varied viewpoints. In such situations, stakeholders use very different terms to describe their objectives, with relatively little overlap in their evaluation concerns, so a single objectives hierarchy is not warranted.

Structuring multiple objectives is also useful for strategic planning within organizations. Commercial companies may, for instance, find that generic objectives such as maximizing profitability, achieving long-term growth, and maintaining security can aid management in decision analyses.

APPENDIX V.
EXAMPLE DECISION MAKING CHECKLISTS

Checklists can help to improve the quality of decisions. Table 9 provides an example of a post-decision review questionnaire that can be used to evaluate how decisions were made and capture lessons learned. Table 10 provides an example of questions to ask via a checklist that can assist in reviewing a DM process from the context of potential biases.

TABLE 9. DECISION DIAGNOSTIC QUESTIONNAIRE (ADAPTED FROM [31])

Question	Response
Were the decisions right?	
Were the decisions made with appropriate speed?	
Were the decisions executed well?	
Were the right people involved in the right way?	

Was it clear for each decision:
   – Who would recommend a solution?
   – Who would provide input?
   – Who had the final say?
   – Who would be responsible for following through?	

Were the decision roles, process, and time frame respected?	
Were the decisions based on appropriate facts?	
To the extent that there were divergent facts or opinions, was it clear who made the decision (formal decision maker)?	
Were the decision makers at the appropriate level in the company?	
Did the organization's measures and incentives encourage the people involved to make the right decisions?	

TABLE 10. CHECKLIST: QUESTIONNAIRE ON BIASES IN DECISION MAKING (ADAPTED FROM [31])

Check for	Description
Self-interested bias	Is there evidence that self-interest may impact or influence the decision?
Heuristics effects	Is there evidence of any bias in the decision; has the team fallen in love with its proposals?
Groupthink	Have dissenting opinions been considered in the decision? Has a "black hat" or devil's advocate been used to challenge the decision?
Saliency bias	Could the decision be overly influenced by an analogy to a memorable success?
Confirmation bias	Are credible alternatives included along with the recommendation?
Availability bias	If you had to make this decision again in a year, what information would you want, and can you get more of it now?
Anchoring bias	Has the decision used unsubstantiated information extrapolated from history?
Halo effect	Is the team assuming that a person, organization, or approach that is successful in one area will be just as successful in another?
Sunk cost fallacy and endowment effect	Are the recommenders overly attached to a history of past decisions?
Overconfidence fallacy, optimistic biases, and completion neglect	Is the base case overly optimistic?
Disaster neglect	Has the worst case been considered, and is it bad enough?
Loss aversion	Is the recommending team overly cautious?

APPENDIX VI.
VALUE-IMPACT ANALYSIS

Below are two examples to illustrate value-impact analysis that would help DM.

VI.1.   ALARA

According to International Commission on Radiological Protection (ICRP) Publications [ICRP Publication 9 [76], para.52, ICRP Publication 22 [77], para.11]:

- "Any unnecessary radiation exposure be avoided, and all doses be kept as low as is readily achievable, economic and social considerations being taken into account";
- "It is possible to define the point at which it can be said that a dose is as low as is readily achievable, economic and social considerations being taken into account, by choosing the dose at which the economic and social gains of further reducing the dose are equal to the economic and social costs of achieving that reduction".

The justification of this dose level by the expected benefits requires the use of cost-benefit analysis, and this analysis (Appendix II of Publication 22) defines the method to convert radiation exposure to monetary values.

The optimization of radiation protection is also defined in ICRP Publication 103 (2007) [110] as the process by which the likelihood of exposure, the number of people exposed, and the magnitude of individual doses should be kept as low as reasonably achievable, taking into account economic and societal factors (p.14). This concept is considered synonymous with ALARA.

Furthermore, ICRP Publication 101b (2006) [111] positions optimization not merely as dose reduction, but as a broad decision making process that includes stakeholder involvement and the promotion of safety culture.

USNRC's 10CFR50 Appendix I gives a Numerical Guide for Design Objectives and Limiting Conditions for Operation to Meet the Criterion "As Low as is Reasonably Achievable" for Radioactive Material in LWR Nuclear Power Reactor Effluents.

VI.2.   BACKFITTING AND DEPENDENCY OF BALANCING COST ON "ADEQUATE PROTECTION"

By rule (10CFR§50.109), the USNRC is mandated to justify imposing backfit requirements on licensees by a cost-benefit analysis (CBA). The increased public protection due to the backfit, expressed in averted radiation exposure, is compared to its implementation costs. A recent example of cost-benefit analysis for backfitting is imposing "severe accident capable vents" to BWRs with Mark I and II containment in 2013. The initial NRC requirement established in 2012 was for venting to prevent core damage in case residual heat removal capability is lost. However, this was overridden because the vent could operate during severe accidents. The NRC had concluded that:

(1) "the requirement to provide a reliable hardened vent to prevent or limit core damage upon loss of heat removal capability is necessary to ensure reasonable assurance of adequate protection of public health and safety and

(2) the requirement that the reliable hardened vent remain functional during severe accident conditions is a cost-justified substantial safety improvement."

In a cost-benefit analysis of the above (1), the conversion of estimated dose savings into dollars used the following equation [79] :

[(Estimated Accident Frequency) x (Change in Population Dose)] x ($2,000/person-rem) x [1-exp (- (discount rate) x (remaining reactor life))]/ (discount rate)

PSA and severe accident analysis codes MELCOR and MACCS supported this dose-saving assessment.

Note that the evaluation of benefit here is only about public health. Traditionally, the USNRC's position in the CBA was to limit the evaluation of benefits to "adequate protection to public health and safety" and not to include offsite property damage such as that caused by land contamination. An important thing to note in this CBA is whether offsite property damage is included in "adequate protection" [112], depending on which the range of cost-justifiable requirements varies.

This discussion regarding nuclear regulation concerns damage to the public since the modification using CBA. However, accident cost has a wider spectrum of attributes that includes liability born by the Owner/Operator, loss of onsite property at NPP and its recovery, decontamination of land, disposal of accident-generated water and other wastes, etc., most of which the responsible Owner/Operator will have to bear. Hence, the scope of damages to be covered in CBA for "adequate protection" by industry would be *much larger* than the case of a regulatory requirement, which means an increased amount of balancing (in CBA analysis) level of investment to avoid accident accompanying large release of radioactivity to the environment. If nuclides with a long half-life for decay, such as Plutonium, are involved significantly (in most cases, Cs-137 is dominant), habitability can be very difficult for an extended period.

APPENDIX VII.
COUNTRY-SPECIFIC EXPERIENCES

VII.1. ARMENIA

VII.1.1. Introduction

For legitimate and accurate DM on modifications directly related to changes in plant configuration (such as changes in operating limits and conditions, designs of systems and components, technological processes, operating procedures, etc.) and modifications of the management system (such as changes in the organizational structure, safety assessment methods, etc.), as well as the implementation of new regulatory and technical documents, the Armenian NPP (ANPP) operates a Scientific and Technical Council (STC). The STC of the ANPP is a permanent advisory body under the Chief Engineer of the ANPP.

The document "Regulations on the Scientific and Technical Council of the ANPP" has been developed and is in effect to regulate the activities of the STC at the ANPP. This document defines the tasks, functions, work procedures, rights and obligations of the members of the STC.

The STC is guided in its activities by:

- The legislation of the Republic of Armenia;
- Norms and rules in force in the field of nuclear energy;
- Orders and instructions of the Ministry of Territorial Administration and Infrastructure of the Republic of Armenia;
- Orders and instructions of the ANPP management;
- IAEA recommendations.

The composition of the STC includes the Chief Engineer of the ANPP as the chair of the STC, the Head of the Engineering and Technical Support Department as the secretary of the STC, deputy chief engineers and some heads of divisions and departments.

VII.1.2. Functions and processes of STC

The main tasks and functions of the STC of the ANPP are:

- To review and decide on issues related to technical policy and strategic planning related to technical issues at the ANPP;
- To organize expert examinations of design, engineering, technological developments aimed at improving safety, the economic operation of the ANPP and make decisions based on the conclusions and proposals of the members of the STC;
- To review and decide on the feasibility and possibility of introducing new technologies and equipment at the ANPP;
- To review and decide on improving operations management and introducing high-tech tools into management;
- To review and accept technical proposals submitted by divisions of the ANPP;
- To review individual issues requiring additional discussion based on the results of verification or validation;
- To review new regulatory documents for implementation at the ANPP;
- To review proposals on the possibility of changes in the structure of divisions.

The STC is managed by the chairperson – Chief Engineer of the ANPP. In the absence of the Chief Engineer, a designated replacement performs the duties of the chairperson of the STC.

The organizational arrangements of the STC meeting preparations are assigned to the secretary. The chairperson convenes meetings of the STC as necessary.

For each meeting, the secretary of the STC prepares a meeting plan. The secretary of the STC prepares materials for consideration at the meeting and sends them to all STC members and invited specialists six days before the meeting.

Meetings of the STC are held if more than 50% of the total number of STC members are present.

Decisions of the STC are made by open voting with a simple majority of votes from the members present. In the event of a tie, the chairperson's vote is decisive.

After the meeting, the secretary of the STC prepares the minutes within four days. The opinions of all participants that differ from the decisions made have to be recorded in the minutes. Two copies of the minutes are made: One is archived, and the secretary of the STC keeps the other.

The minutes are signed by all the members present at the meeting. The secretary of the STC then submits the minutes to the chairperson of the STC for signature.

The decision of the STC is not an administrative document. An order from the Chief Engineer or the Director General of the ANPP is required to implement the decisions taken.

Individual specialists of the ANPP, as well as specialists from other organizations or enterprises, may participate in the STC meetings at the chairperson's invitation but without voting rights.

VII.1.3. Conclusion

Discussions at the STC ensure that decisions are made in a collegial format, considering the opinions of all council members. The decisions made by the STC guarantee the high-quality implementation of all the modifications and modernizations.

VII.2.   FRANCE

VII.2.1. Decision making guide

EDF has issued a guide for its managers, relying on a set of principles to facilitate rigorous decisions, not omitting steps and internal reviews before DM.

The process relies on seven steps (see Fig. 6.), whose key features are:

- These principles are neither exhaustive nor restrictive and can be adjusted according to the assignment of the operational entity (including engineering activities) or context;
- Personnel are encouraged to have a careful and questioning attitude;
- Skilled staff are engaged to support technical or organizational decisions;
- Decision makers are skilled in understanding problems and have the right posture to listen to various points of view;
- Leaders, managers, and operational staff have to stick to their respective roles in DM, which are explicitly described and defined before the DM;

— Open dialogue is encouraged when discussing decisions, and contradictory arguments are recognized and promoted.

Figure 6. provides traps and advice identified at the different steps of the DM process.

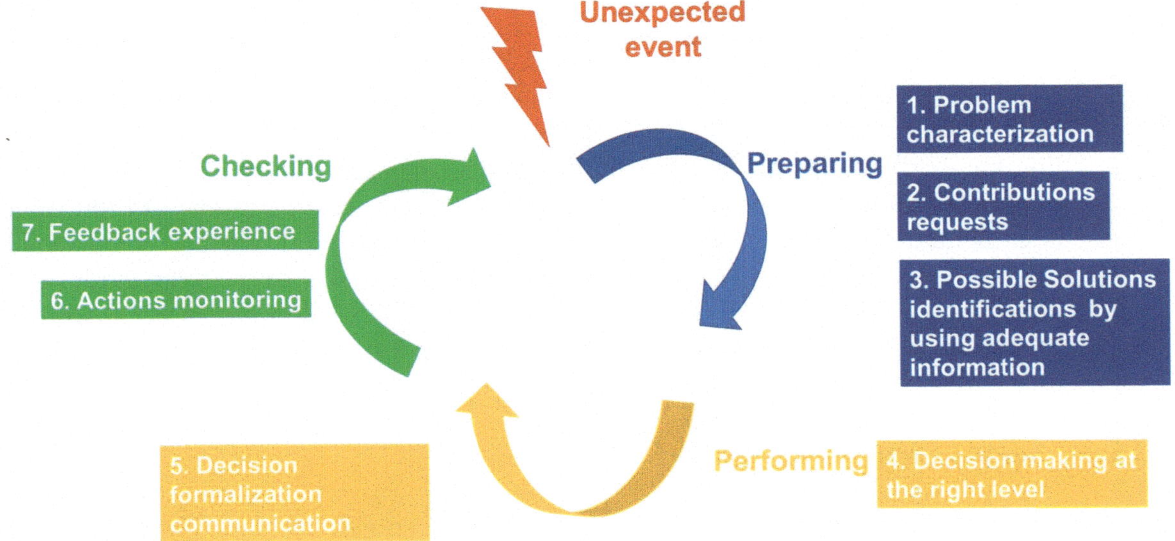

PREPARATION WITHOUT CONSIDERING RELEVANT FEEDBACK

☹ Only the information confirming the exposed point of view are shared. The problems are mixed up
☺ Search for information beyond the starting point of view
☺ Open a wider vision

DECISION STEPS INTO THE ANALYSIS

☹ Biases are introduced in the identification of solutions if DM is involved in the analysis.
☺ The decision maker steps back and leaves the teams free to identify different solutions

LACK OF EXPERTIZED SUPPORTIVE STAFF

☹ Risk of not addressing an issue which is nevertheless needed
☺ Point out the issues and direct & indirect problems
☺ Check that staff are able to answer to each listed point

POOR QUALITY OF AVAILABLE INFORMATION

☹ Information is partial or oriented
☹ A limited number of sources has been used
☺ Accurate search for information which may contradict facts is performed
☺ Various sources of information is used

POORLY STATED PROBLEM

☹ It refers directly to the solution
☹ The situation is insufficiently described
☹ Various problems are mixed up
☺ Review what is observed & bothers
☺ Point to the problem by a shared analysis between the main actors / contributors.
☺ Prioritize the problems

FIG.6. EDF Decision Making Guide, Traps and Advice (courtesy EDF)

VII.2.2. Design authority

EDF's DA is organized under a formal team with engineers who monitor design changes at different levels:

- For future design evolutions, the DA checks if the decisions within the projects are consistent with the result of safety studies and the design changes meet the updated safety requirements;
- For the NPP modifications, the DA issues formal reports checking the adequate implementation schedule of the safety improvement modifications and the related documentation changes (Safety Analysis Report and General Operating Rules). This report, named "Design ID Card," is produced for every NPP of the fleet and for every main outage (intermediate and decennial outage). This reporting process allows a second-level control, in addition to the process of modifications implementation on the NPPs.

VII.2.3. Supply Chain

Some experience in DM related to supply chain in the EDF shows:

- Supply chain impacts on organizational and technical choices:

  - Close collaboration between EDF as a decision maker and Framatome as a supplier allowed for the development of an innovative solution for working on bottom head instrumentation tubes. The solution took advantage of the vendor's knowledge of this type of operation for vessel heads ("water jet peening").
  - Interface optimization between contractors and subcontractors, based on the contract of "turnkey services" by subcontractors, allowed effective implementation of works during outages. This was the case, for instance, in the pressurizer heater replacements: the subcontractor was in charge of all the activities including logistics tasks (scaffoldings). Another example is the handling of crane start-up operation: and the related spare parts were made available to the subcontractor in order to optimize outage sequences.

- Supply chain impacts on industrial programs strategies:

  - Contracts may be decided to maintain skilled industrial partners in the long term. Some activities are planned to have repetitive operations performed by the same subcontractors. Within one outage organization, maintenance slots are adjusted to consider specialized resource availability. This adjustment can involve splitting important operations between various outages when specific skilled teams are required. Conversely, contracts can be established to ensure enough staff are available to perform 3x8h shift activities (e.g., for steam generators and tube plugging).

- Supply chain impacts on the supervision of activities on site:

- For pressurizer valves maintenance, opening/closing of vessels, diesel maintenance, and critical phases have been identified with the contractors, the NPP staff sets up checkpoints, and orientation and training systems are put in place to guarantee a high-level quality on these operations.

VII.2.4. Artificial Intelligence

Innovation and progress in numerical tools and increased digitalization of processes may be considered to support DM.

EDF has developed a tool called "Metroscope" which couples a digital twin of the NPP with AI to detect component problems in an anticipated way, which may impact plant performance.

Three components are part of the device (Fig. 7.):

- The real process within the plant, which is equipped with hundreds of points of instrumentation that capture and transmit plant information;
- The digital twin of the process which generates simulated data. The program simulates every plant component and its relations with other components and uses the physical equations to predict the plant behaviour. The existing defaults of the NPP are also built in;
- The AI compares the real measures and the simulated data. A discrepancy between the data may reveal a potential issue: Artificial Intelligence studies all possible scenarios to diagnose their origin.

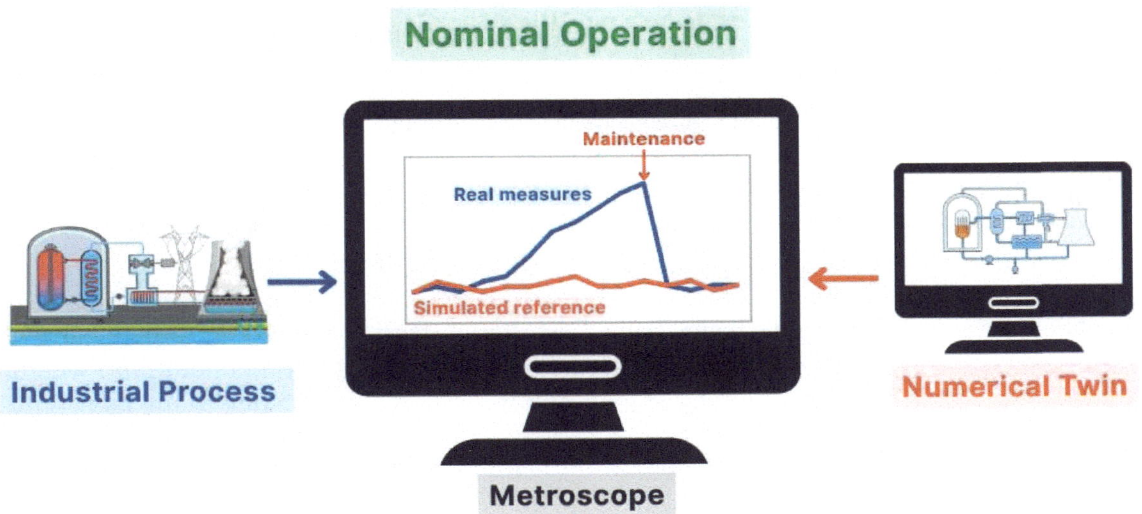

FIG.7. EDF Metroscope AI System (Courtesy EDF)

The system relies on Bayesian networks, which identify the causes, starting from the effects produced: if a physical defect is characterized by a specific impact on a measured parameter, then when AI generates this impact on the parameter, the cause has been identified through the simulated scenario.

VII.3.1.   Method for structured decision making – FOR-DEC

In the early 1990s, the German aerospace research and technology centre, "DLR", developed a method for structured decision making named FOR-DEC [117]. Nowadays, it is widely used in the aviation and medical industries. German nuclear operators use the tool for operational DM, but it can also be very helpful for strategic and tactical DM.

FOR-DEC is an acronym for the following steps of DM (Fig. 8.):

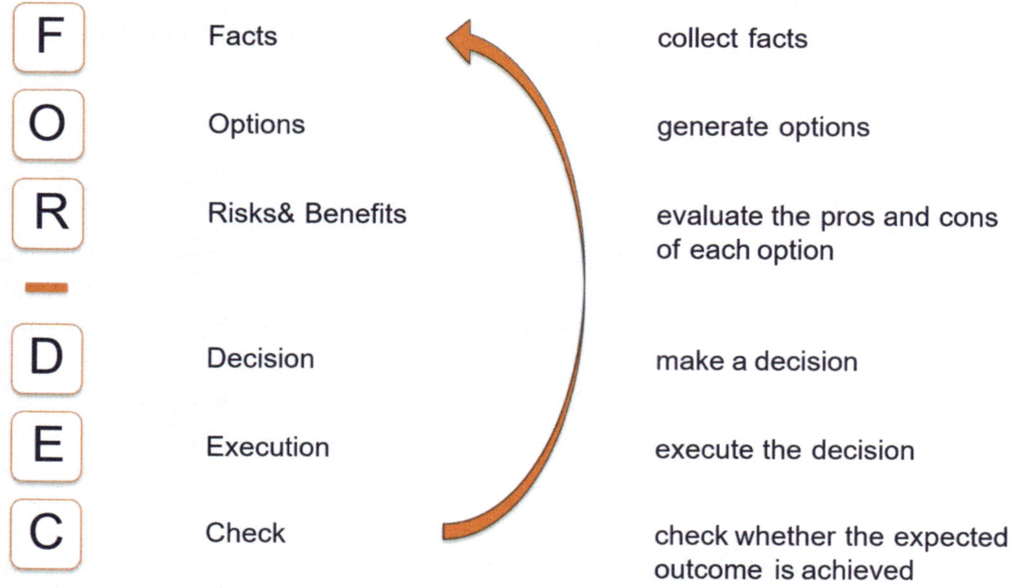

FIG. 8. The FOR-DEC model

The different steps are executed chronologically. By strictly following the order, this method offers well-founded, efficient and transparent decisions. Some details on the various steps are below. Only after one step is executed completely, start with the next one.

F: Facts	The goal is to collect as many facts as possible. Make sure that everybody involved in the process is heard. At this point, no weighing of facts is made, and no options are discussed. Stick to the facts only.
O: Options	Generate a reasonable number of options. Do not stop after the first most obvious options, and do not rate or rank the options. This step can be very creative and open-minded, and it does not refuse options that might not sound beneficial initially.
R: Risks & Benefits	Now, all the options have to be evaluated. Name risks and benefits for all options. Do not stop after the first option that looks beneficial. Do not forget to define the risks of the options you prefer and the benefits of the options you do not like.

-	Make a short break and reflect on the discussion. Has anything been missed? You might even want to leave the room briefly to get some (mental and physical) distance.
D: Decision	The decision maker (an individual or a group) has to decide now. This decision has to be communicated and explained.
E: Execution	Tasks have to be assigned. Do not start this before the final decision is made. Ensure everybody understands the decision completely and knows what to do now.
C: Check	Goals to be reached, milestones, etc., have to be defined. Check the outcome of the decision. If it is unsatisfactory, start the FOR-DEC process again by collecting facts.

Does FOR-DEC need much time?

Yes and no. Finding the perfect solution in a complex and dynamic situation will take time, sometimes weeks or months. Obviously, if time is not critical, FOR-DEC can be used. It helps structure the process and allows the possibility to document the results of the different steps. This work can be very helpful if persons join the process at a later stage and need to be informed. It is also very helpful if a decision has to be explained (or even justified) afterwards.

However, even if the situation is time-critical, FOR-DEC can be used. But now, the perfect solution is not the target, and any option that offers a safe action plan is acceptable. Therefore, the generation of options will be stopped even after the first suitable option is found. No further time is spent searching for more options, and the "R" (Risks and Benefits) has to be evaluated only for one option. This method offers a structured approach even under time pressure and minimizes the risk of missing important facts. As it is true for every decision, start with defining the available timeframe.

FOR-DEC makes a decision transparent if the considered facts, options, risks & benefits are documented and communicated.

VII.4.   REPUBLIC OF KOREA

VII.4.1. Decision making by E-TOWER

VII.4.1.1.   Introduction

Since 1971, the Republic of Korea has embarked on nuclear power plant construction, commencing the first commercial operation of Kori Unit 1 (587 MWe) in 1978. As of December 2024, the Republic of Korea operates 26 nuclear power units with a total installed capacity of 26.05 GWe. The country's nuclear power sector is managed by the government-owned corporation Korea Hydro and Nuclear Power (KHNP). The average operational lifespan of the nuclear plants is 24 years, comprising 12 units of OPR-1000, four units of APR-1400, three units of CANDU, five units of Westinghouse PWR, and two Framatome reactors. The Korean fleet includes three pressurized heavy water reactors, 23 pressurized light water reactors, and four exported reactors to the UAE's Barakah Nuclear Power Plants that adopted the APR-1400 design.

To enhance the safety and operational performance of its diverse fleet, which includes a variety of design backgrounds, reactor types, and operational lifespans, KHNP has developed and implemented an advanced operational management system known as E-Tower. This system is critical in operational, tactical, and strategic DM. E-Tower provides optimized operational and engineering solutions based on equipment status and supports consistent DM across all relevant departments. It ensures strategic and operational directive alignment among the CEO and all executives within the organization.

VII.4.1.2. E-Tower System

[Main Function and Configuration]

This integrated monitoring and management system represents a significant advancement in nuclear fleet operations through three primary functional areas:

- Comprehensive Situation Management
  The system enables real-time monitoring of approximately 1,200 operational variables per unit across all plants. This continuous online diagnostics system provides detailed oversight of key equipment status. The monitoring is conducted independently and parallel to Main Control Room (MCR) operations, establishing redundant surveillance that enhances overall plant safety through dual independent monitoring streams.

- Proactive Action Through Advanced Analytics
  The E-Tower employs sophisticated pattern analysis and detection systems for abnormal signal identification. This capability enables early detection of potential failure signs, allowing for proactive intervention before issues impact plant operations. The system facilitates in-depth, high-precision analysis of operational data, with findings shared with MCR personnel for comprehensive root cause analysis. This proactive approach has proven effective in preventing reactor trips and unexpected plant shutdowns.

- Emergency Response Coordination
  The system provides robust emergency situation control and support capabilities, ensuring efficient communication with regulatory authorities and rapid conveyance of executive decisions. This function is crucial for managing potential scenarios involving radiation exposure or security concerns. The E-Tower is a central hub for information sharing and coordination among all facilities during emergency responses.

The monitoring process involves various control and protection systems that collect operational data from the nuclear power plants. This information is processed through multiple servers and transmitted to the E-Tower, where artificial intelligence-powered early warning systems (EWS) provide independent monitoring and analysis.

When the E-Tower detects an early warning signal, it initiates a systematic response process consisting of six key steps:

- Continuous Monitoring: The system maintains constant surveillance of plant parameters;
- EWS Alarm Generation: When anomalies are detected, the system generates an early warning;
- Initial Analysis: Evaluation of the alarm cause through equipment status and

procedure reviews;

- Detailed Investigation: In-depth analysis incorporating failure history and operational experience;
- Maintenance Coordination: Organization of necessary maintenance activities with relevant departments;
- Knowledge Sharing: Distribution of the incident as a 'one-day briefing' across the organization to prevent recurrence

This systematic approach ensures that potential issues are detected and resolved promptly and serve as learning opportunities for the entire organization. The process is illustrated in Figure 9.

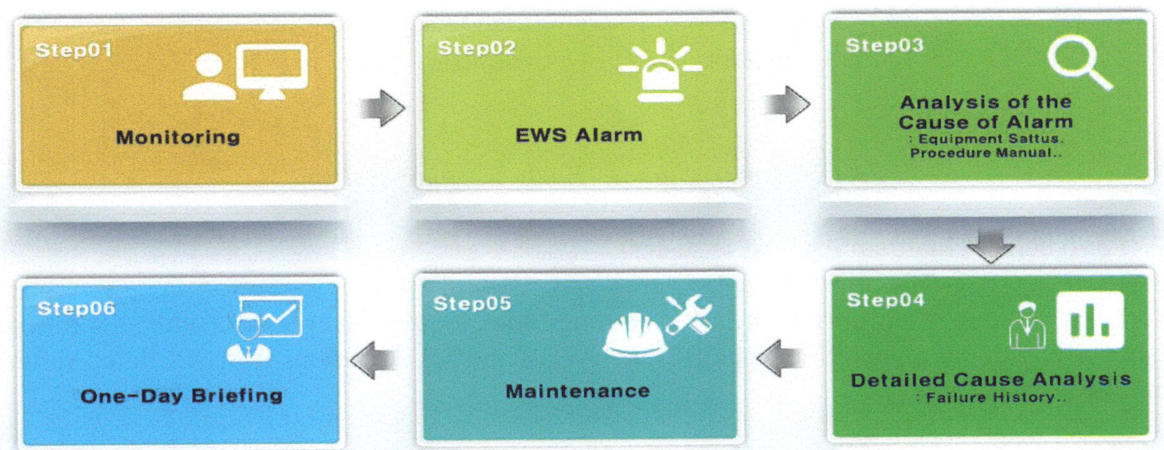

FIG. 9. E-Tower's systematic response process

[E-Tower's Advanced Early Warning System (EWS)]

The E-Tower employs an innovative AI-based early warning system called Smart Signal that enhances traditional monitoring capabilities. This system operates by learning normal operational patterns from historical data to predict the expected behaviour of multiple variables simultaneously.

The key advantage of this system lies in its ability to detect anomalies before they trigger conventional MCR alarms. While traditional alarm systems rely on fixed thresholds, the EWS monitors dynamic trends in operational parameters. By comparing real-time data against learned patterns, the system can identify deviations from expected behaviour much earlier, providing operators with additional response time, as shown in Figure 10.

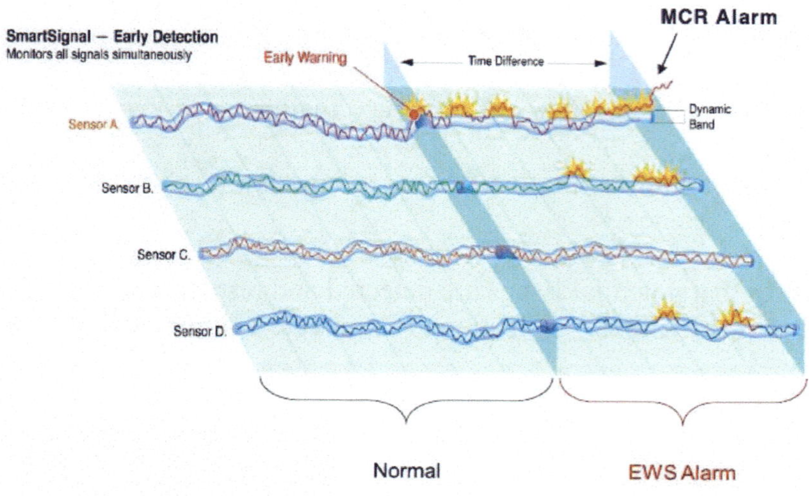

FIG 10. Smart Signal - Early warning system

[EWI & Nuclear Guardian Against Disaster]

KHNP has developed a proprietary Early Warning Index to provide operators with an intuitive understanding of plant operational status. The EWI quantifies deviations from predicted values using the formula:

EWI (%) = |Present Value - Predicted Value| / ((Present Value max - Present Value min) × M factor) × 100

Where the M factor is a sensitivity adjustment factor. The system presents this information through a colour-coded visualization system with four levels:

- Stable (Green)
- Attention (Yellow)
- Caution (Orange)
- Warning (Red)

This visual representation allows operators to quickly assess the operational status and identify potential issues before they become critical.

[Nuclear Guardian Against Disaster (NGAD)]

The E-Tower incorporates a comprehensive disaster monitoring system called NGAD, which integrates external data from national agencies to monitor and assess natural disaster risks to nuclear facilities. The system provides:

- Real-time monitoring of natural disasters, including forest fires, typhoons, and earthquakes;
- Integration of data from the Korea Meteorological Administration and Korea Forest Service;
- Predictive analysis of disaster impacts on multiple facilities;

– Support for executive decision making during multi-site events

For forest fires, the system tracks outbreak information and predicts spread patterns. For typhoons, it analyses historical patterns to simulate potential impacts on NPPs. This integrated approach enables proactive measures and preparedness across the entire fleet of nuclear facilities.

VII.4.1.3. *Outcomes and Expectation*

The E-Tower system has a wide range of impacts on NPP operations:

– Enhanced Safety and Reliability:

E-Tower enables real-time monitoring and analysis of safety indicators, ensuring swift and reliability of nuclear power operations.

– Optimized Decision Making:

E-Tower facilitates informed DM at the operational, tactical, and strategic levels by integrating operational data and providing actionable insights. It ensures that decisions are consistent and aligned with organizational goals.

– Improved Operational Efficiency:

The system provides tailored engineering solutions and operational guidelines based on the real-time status of facilities. The system outputs lead to more efficient use of resources and minimizing downtime.

– Effective Coordination:

E-Tower fosters seamless communication across all departments, from plant-level teams to executive management. This unified approach ensures that all stakeholders have access to the same critical information, streamlining collaboration and follow-up actions.

– Proactive Maintenance:

With its advanced analytics capabilities, E-Tower predicts equipment failures and schedules maintenance proactively, reducing the likelihood of unplanned outages and extending the operational lifespan of facilities.

– Strategic Alignment:

The system supports long-term strategic planning by analyzing trends and providing insights into operational performance. These insights ensure that the organization remains aligned with actions needed to ensure safety and the operational lifespan of facilities.

– Transparency and Accountability:

> E-Tower promotes a culture of transparency by sharing comprehensive operational reports with all departments and executives. This transparency enables accountability and continuous improvement in processes.

Through these impacts, E-Tower enhances the safety and performance of NPP operations and strengthens KHNP's position as a global leader in nuclear energy management.

[A Case Study: Gyeongju Earthquake in 2016]

The E-Tower's capabilities were notably demonstrated during the 2016 Gyeongju earthquake sequence, showcasing its effectiveness as a centralized control and decision making hub for KHNP's nuclear fleet. The incident began on September 12, 2016, with a magnitude 5.1 earthquake at 19:44, followed by a stronger magnitude 5.8 earthquake at 20:32. The seismic activity continued, with 423 recorded earthquakes in the region through September 22.

The E-Tower facilitated coordinated action across multiple facilities in response to this unprecedented seismic event. The system's real-time monitoring capabilities and integrated disaster management functions enabled KHNP to implement a systematic response protocol. This protocol included the sequential manual shutdown of Wolsong Units 1-4, executed according to emergency operation procedures when seismic measurements approached the operating basis earthquake threshold of 0.1g.

The emergency response was executed with precision timing: Unit 1 at 23:56, Unit 2 at 00:04, Unit 3 at 00:10, and Unit 4 at 00:15. Following the shutdowns, comprehensive safety inspections were conducted from September 13 to 18, involving approximately 350 personnel from headquarters, power plants, and partner companies. These inspections focused particularly on equipment susceptible to seismic effects, such as EDG fuel tanks, transformers, and pumps.

The E-Tower's effectiveness during this crisis was attributed to two key factors: its comprehensive real-time monitoring capabilities, providing crucial data for decision making, and its advanced AI-based support systems, including the Early Warning System (EWS) and Nuclear Guardian Against Disaster (NGAD). The seamless information sharing between headquarters and individual plants and the management's ability to make swift, informed decisions ensured stable operations throughout the seismic event.

This real-world demonstration of the E-Tower's capabilities validates KHNP's integrated approach to nuclear fleet management, particularly during natural disasters affecting multiple facilities simultaneously. The system's performance during the Gyeongju earthquake is a testament to the value of centralized, technology-driven oversight in NPP operations.

VII.4.1.4. Conclusion

The E-Tower system effectively integrates the three core elements of nuclear power operations: plant, process, and people, to support DM processes comprehensively. By leveraging real-time data alongside historical data from the past 18 months, E-Tower provides an early warning system and status indicators (M Factor) that enable precise technical decision making at the plant level. Additionally, the system supports headquarters-level nuclear management and strategic planning.

Since its implementation, KHNP has achieved significant advancements, including:

- Minimized Decision Making Errors: E-Tower ensures more accurate and reliable decisions by integrating robust data-driven insights;
- Real-Time Responsiveness: The system enables world-class safety operations through immediate and effective responses to operational conditions;
- Regulatory and Policy Compliance: It offers extensive information to address regulatory requirements and align with national policies;
- Public Trust Enhancement: E-Tower's contributions to safety and transparency have significantly bolstered public trust in nuclear energy operations.

Furthermore, E-Tower is continuously upgraded with AI functionalities, enhancing its analytical and predictive capabilities. KHNP also explores opportunities to expand the system's technological foundation to international NPPs, fostering global collaboration and knowledge sharing. This strategic development positions KHNP as a global leader in innovative and safe nuclear power management.

VII.5.   RUSSIAN FEDERATION

The Russian operating organization Rosenergoatom has created and is consistently improving the Decision Making Centre (DMC).

The Decision Making Centre is a platform for collecting, storing and processing operational data and a space for Rosenergoatom top-management meetings and decision making.

The DMC is located at the Rosenergoatom headquarters and is well-equipped for the work of managers and specialists of the organization. The most valuable part of the DMC is the software, which is the basis of DMC. Programmers worked closely with future users during its development, using their experience and considering user's suggestions.

The DMC was created at the initiative of sales departments, and the initial functionality was focused on supporting decision making when working at the wholesale electricity and capacity market (WECM). Working at the WECM is the final stage of the technological chain of energy companies, but its rates are indicators of the economic efficiency of a company.

The objective of an operating organization is to support NPP operations and ensure the safety and reliability of NPP equipment. At the same time, it is important to effectively sell electricity at the WECM and minimize the risk of loss in case of deviations. Production departments and commercial dispatchers have to receive technological and market data promptly for that purpose.

As a result of development, the DMC was adapted for use in all business areas of Rosenergoatom. The Centre was involved in Rosenergoatom's daily activities. Today, it enables managers to make prompt and accurate decisions and avoid paper reports and additional requests from experts.

Fifteen systems are integrated into the DMC, which monitor more than two thousand real-time indicators from hundreds of sources. Most of the indicators are collected automatically. Data comes from all Rosenergoatom NPPs, including construction sites and infrastructure organizations of the energy system and market.

Since the launch of the DMC, all information is concentrated in a single system. Modern means of communication, multimedia equipment, and a set of dashboard panels consisting of widgets with key activity indicators allow a person to perform a detailed analysis of collected data. Real-time data support reports and proposals for decisions during top-management meetings.

An AI digital assistant is involved in DM. It instantly notifies users of every change in the operation of NPP units, calculates the impact of an incident and automatically sends results to a manager. The introduction of AI allows using voice commands to control the DMC.

The speed of decision making is very important for operating organizations. For example, from a financial perspective, the cost of a possible optimization could be millions of dollars. With DMC tools, users receive structured data and analytics, are instantly informed on changes in operation, and perform rapid calculations. The DMC enables contacting NPP management as a matter of priority, either to make a decision or to clarify the situation, and inform Rosenergoatom management. In the event of NPP equipment shutdown, the DMC calculates the impact, including financial, and could simulate an action scenario.

The DMC is a step towards the digital management of the industry based on effective collection and processing of data, monitoring, forecasting and calculations, and supports prompt decision making.

Digitalization and automation of calculations and interaction between Rosenergoatom departments improved managerial decisions to a fundamentally new level. Some business processes have accelerated from two or three days to three or four minutes. An example of optimized processes is electricity generation and revenue forecasting, whereby a user receives an updated forecast in real-time in case of changes in power equipment maintenance schedules.

Many crucial strategic and tactical decisions have been made in the DMC area.

VII.6.   USA

VII.6.1. Decision Gates for Advanced Nuclear

VII.6.1.1.   Background

The Tennessee Valley Authority (TVA) is a federal corporation wholly owned by the United States government. TVA operates the nation's largest public power utility, with seven large light-water reactors located across three sites. TVA has a strategic goal to identify energy innovations, including potential new nuclear options, to provide safe, reliable, and clean power generation for the Tennessee Valley.

In 2022, the TVA Board of Directors announced a "New Nuclear Program" to help develop nuclear generation options by the early 2030s. The program focused on the first potential SMR demonstration at TVA's Clinch River Nuclear site and the potential to deploy multiple nuclear technologies at multiple sites within the service territory.

With a service territory crossing seven states and over 10 million customers, TVA actively manages multiple construction projects worth approximately $900M annually. With this construction experience comes a well-established project management organization and oversight framework which meets or exceeds industry norms. In considering the potential for new nuclear deployments, TVA decided that an additional layer of communication and decision making would be appropriate, given potential project costs and interest in nuclear power.

TVA developed a three-tier "Decision Gate" (DG) process to provide for structured executive leadership engagement at the start of each new phase of work. The Decision Gate framework offers strategic opportunities for executive leaders to weigh in on the progress and efficacy of the project before the organization proceeds with the next phase of work. A standard set of DG criteria was created to ensure a broad set of enterprise considerations will be evaluated and presented to support decision making. Following an executive review of the DG criteria reports, a final recommendation is made to the Board of Directors regarding how to proceed at each gate.

VII.6.1.2. Decision Gate (DG) Process

<u>Overview</u>

TVA's phased "Decision Gate" process divides the preparations for a potential nuclear plant deployment into three phases: Planning, Project, and Construction. A DG precedes each phase of work and allows executive leadership to assess progress and specifically authorize moving into the next phase of work.

The progressive phases of work in the DG process reflect a growing commitment toward deployment while still affording an opportunity to proceed, delay, or off-ramp as appropriate to manage risk and other enterprise considerations. (See Figure 11.).

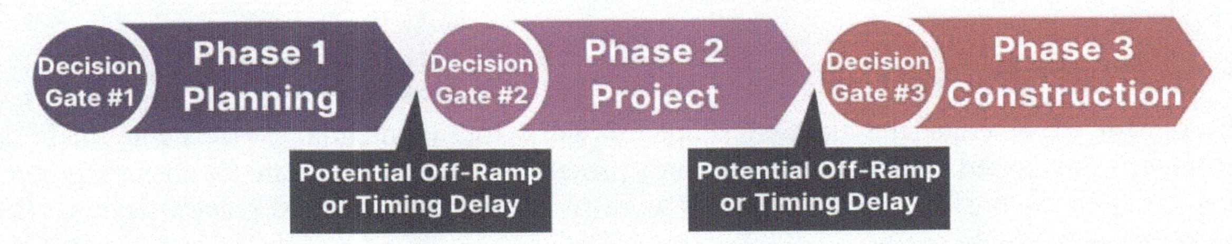

FIG. 11. Decision Gate Process

Six criteria are evaluated at each DG, and the critical thinking for each aspect of the DG recommendation is documented. The criteria are intended to capture a broad range of enterprise considerations for new nuclear deployment. The criteria were selected to provide sufficient breadth to encompass the evolving level of detail expected at each DG and include the following areas: Technology Readiness, Planning Readiness, Cost Estimates, Schedule Estimates, Partnering Readiness, and Enterprise Fit.

A Criteria Development Team (CDT) was established, with representation from each relevant enterprise group, to identify and reach a consensus on the specific measures for each criterion. While the CDT established the criteria, cross-functional teams best suited to the topical area developed the material supporting each decision gate criteria. These packages are then subjected to a multi-business unit review for concurrence. An Executive Leadership Team sponsor further reviewed and approved the specific criteria. The individual criteria are described below.

<u>Decision Gate Criterion Definitions</u>

The criterion definitions are expected to expand with each DG as the level of planning and information available to assess progress in each area increases. The examples below reflect the Criterion at DG1 for a first-of-a-kind (FOAK) advanced reactor design with conceptual design information.

(Technology Readiness Criterion at DG1)

Identification of the developing nuclear technology that is likely best positioned to facilitate goals to have a reliable, affordable, flexible, and clean advanced reactor option available by the 2030s. This review includes whether the technology is sufficiently mature to proceed into a phase of detailed design, scoping, estimating, and planning.

(Planning Readiness Criterion1 at DG1)

Evaluation of whether sufficient pre-planning has been done to ensure the program can quickly move forward and whether the planning risk profile is acceptable. Assessment has to consider factors such as working with experienced industry partners and subject matter experts to reduce the risks of a new reactor design and construction. Best practices for reactor design and construction need to be considered early.

(Cost Estimates Criterion at DG1)

Assessment of the project's defined scope of work as an input to prepare a cost estimate and the cost estimate itself to determine whether proceeding into Phase 1 work and detailed scoping, estimating, and planning (DESP) is justified. The cost estimate maturity (AACE Class 5 to Class 4) and whether it satisfies industry best practices for a high-quality cost estimate are considered.

(Schedule Estimates Criterion at DG1)

Assessment of whether the defined scope of work for a potential SMR deployment is sufficiently developed to prepare a preliminary project schedule appropriate for an early project stage. Factors to consider are the scope of work for developing standard reactor design, site-specific design, and licensing.

(Partnership Assessment Criterion at DG1)

Assessment of the strategy to establish partnerships and whether sufficient potential for collaboration exists to justify moving into the next phase of work. Considerations include whether a like-minded group of participants (prospective partners) is available and supports similar goals for new nuclear technology development (e.g., commercialization by 2030s with cost-competitive technology). Finally, an assessment of whether prospective partners possess financial, technical, and market attributes compatible with the project objectives.

(Enterprise Fit Criterion at DG1)

Assessment of how the proposed technology can provide potential value relative to the company's long-term generation asset strategy and decarbonization goals. The assessment has to also consider whether the potential value, as currently understood, warrants the level of investment needed to advance to the next DG.

In Phase 1, the preliminary work on design and DSEP done before DG1 is expanded to provide more rigorous reviews for all aspects of the project. The conduct of these reviews is shifted from that of an SMR development team to that of a project management organization. Project management best practices are applied to ensure the development of more comprehensive assessments to support the decision at DG2.

VII.6.1.3. Detailed Scoping Estimating and Planning (DSEP) Process

During Phase 1, the team developed a detailed scoping, estimating, and planning (DSEP) process to support more refined inputs supporting DG2 and DG3 decisions with detailed assessments using industry standards (e.g., AACE cost estimating guidance).

The DSEP process has 3 phases: Baseline to document current status and initial gap assessment, Phase 1 to support Decision Gate 2 for project authorization, and Phase 2 to support Decision Gate 3 for construction authorization. The duration and scale of nuclear projects combined with TVA generation planning call for early DSEP work to inform long-range enterprise decisions with the appropriate rigour and detail.

The scope of the DSEP packages in Phase 1 increased significantly, encompassing the original DG criteria and adding granularity on topics such as design maturity, suitability for TVA's generation needs, and construction planning.

DSEP Purpose

DSEP is a systematic scoping, estimating, and planning process to support project decision making. TVA can make informed project decisions by creating DSEP packages that cover various scopes of work and other topics that affect decision gate criteria.

There are three main purposes of DSEP:

- To Document Scoping and Estimating - DSEP packages will document the scope and estimate, along with the basis and confidence thereof, for various project scopes. Transparency and sufficient detail of scopes and estimates are required to support decision gate criteria information needs;
- To Inform TVA Analysis and Decision Making - DSEP packages will provide the information needed to fulfil decision gate or other key milestone criteria;
- To Identify Work Needed to Meet Decision Gates - DSEP packages will identify criteria and information gaps needed to fulfil decision gate or other key milestone criteria. Gap closure plans will create a successful pathway to bridge the current scope of work to satisfy the decision gate criteria and information needs.

DSEP Scope Changes Based On Work Phases

Early identification of gaps and closure plans will enable the team to efficiently allocate the right resources to inform project decision gates successfully.

DSEP work is organized into three main phases (Baseline, DSEP Phase 1, and DSEP Phase 2) based on the increasing detail and rigour needed to inform the DGs for progressive investment.

The baseline is focused on developing the DSEP process and preparing initial DSEP packages. The baseline DSEP packages assess the status of topics addressing the basis of scope and estimates. Information gaps will be identified, assessed, resolved, and tracked.

Phase 1 packages will address specific criteria and have a direct input to DG 2. Inputs to DSEP packages in phase 1 will include design and licensing deliverables, the closure of many major key project decisions, and formal project estimates and implementation plans.

Phase 2 packages will address specific criteria and have direct inputs to DG 3. Phase 2 will update the DSEP packages with more mature and detailed information.

VII.6.1.4. Conclusion

The Decision Gate process is a critical component of TVA's New Nuclear Program. It enables the organization to make strategic, risk informed decisions as it explores developing and deploying advanced nuclear technologies. TVA can effectively manage the inherent risks and complexities of first-of-a-kind nuclear projects by implementing a structured, multi-phase approach with comprehensive evaluation criteria. Importantly, the DG process is supported by thorough pre-construction planning through the DSEP framework, helping to ensure TVA is well-prepared to execute successful nuclear projects that deliver on schedule and within budget.

VII.6.2. Risk Informed approach and Artificial Intelligence applications

The current key US industry-wide initiatives being implemented to improve DM are risk informed completion times (RICT) of technical specification (TS) and USNRC 10 CFR 50.69, "Risk-Informed Categorization and Treatment of Structures, Systems and Components for Nuclear Power Reactors" [113] rule. In addition, initial plans are being developed to implement AI applications in targeted areas that do not compromise safety.

The PWR Owners Group (PWROG), in partnership with the BWR Owners Group (BWROG), have developed NRC-approved changes to plant standard technical specifications that allow the use of the plant-specific PSA models to determine alternative TS equipment completion times considering the actual plant equipment configuration. Supporting documents and guidance to implement the 10 CFR 50.69 [113] rule efficiently have been developed, and initial areas for strategically applying artificial intelligence applications are being discussed with the USNRC.

VII.6.2.1. Risk Informed TS Completion Times

Implementing RICT of TS uses the plant-specific PSA model to assess current plant equipment configuration and TS equipment modelled in PSA. US experience has seen a negligible increase in overall annual plant PSA risk using this technology.

An example of TS allowing for the use of RICT is shown in Figure 12.

B. One required group of pressurizer heaters inoperable.	B.1	Restore required group of pressurizer heaters to OPERABLE status.	72 hours OR In accordance with the RICT Program

FIG. 12. Risk Informed Completion time Example

The required TS changes are described and justified in PWROG/BWROG developed document TSTF-505, Revision 2, "Provide Risk-Informed Extended Completion Times - RITSTF Initiative 4b" [114] and Nuclear Energy Institute (NEI) Report NEI 06-09 [115].

Applying a RICT is currently allowed for TS completion time extensions of up to 30 days. Using a RICT supports improvements in the planned scheduling of online maintenance of TS equipment. It addresses emergent equipment issues with low plant risk in a reduced plant staff stress environment.

The significant plant benefits for use of the PSA model to determine the TS completion time are:

- The impact on plant risk is scientifically based on industry equipment failure rates, human performance, and best-estimate plant performance;
- Allows significantly improved DM for plant online maintenance;
- Improves plant online reliability (capacity factor);
- The impact on plant risk (with TS equipment out of service) is considered based on the current status of the plant equipment.

VII.6.2.2. 10 CFR 50.69 Risk Informed Categorization

Application of the USNRC rule 10 CFR 50.69 Risk-Informed Categorization [113] uses the PSA model to assess the importance of components in a system (low safety significance, high safety significance). The rule allows safety-related system components to be further separated into low and high safety-significant categories. The high safety significant category components are needed to perform the system safety function. This categorization allows plant resources and costs to strategically focus on components with the highest safety significance.

The process of separating system components into low and high safety-significant categories using the PSA model and independent review by plant system experts is based on the USNRC-approved NEI Report NEI 00-04, "10 CFR 50.69 SSC Categorization Guideline" [116]. Once categorized as low-safety significant, the rule allows alternatives to NRC requirements (quality, maintenance, qualification, inspection, testing) that provide reasonable confidence that the low-safety significant components will perform their design function.

The significant improvements in testing, inspection, and maintenance of plant equipment that can be realized with the use of the 10 CFR 50.69 rule are:

- Supports procurement of commercial-grade equipment;
- Lowers cost (70% reduction average component);
- Significant expansion of equipment purchase options (1 vendor to many);
- Shorter equipment purchase lead times (1 year to weeks or months).

VII.6.2.3. Artificial Intelligence Applications

The advent of tremendous new AI capability, including large language model applications, provides a new opportunity to improve DM capability for nuclear power plants. While it is recognized that barriers will need to be put in place to address the potential for incorrect information being generated by these applications, the step change in the ability to analyse large data sets of information and perform predictive capabilities for new plant scenarios in a rapid timeframe with AI opens many new opportunities for focused information processing to support better human DM.

The areas of initial focused use of AI in the US that have been discussed with the USNRC are:

- Smart search and summary processing tools for plant documentation and plant data systems;

- Probabilistic Safety Assessment improvements with HRA, equipment reliability data, and more realistic core damage and large early release success criteria;

- Augmented accident mitigation strategies;

- Equipment troubleshooting with performance tracking;

- Operator training performance and remediation;

- Optimized preventive maintenance and predictive performance tracking.

In November 2024, the Pacific gas and Electric Company (PG&E) launched the first on-site generative AI solution for the US nuclear sector at the Diablo Canyon NPP using Neutron of the Atomic Canyon's AI-processed search solution. The system integrates the Diablo Canyon document systems using optical character recognition, retrieval-augmented generation and AI-processed search to cut search times for a focused search from hours to seconds. Further progress in the initial AI focus areas is being pursued for US plants over the next few years.

VII.6.2.4. Conclusion

In the US, TS RICT is being pursued to improve operational DM using plant PSA models and USNRC-approved TS changes, providing a more accurate assessment of actual plant online risk, reducing plant risk, and improving plant reliability. The USNRC 10 CFR 50.69 rule is being implemented using PSA models to improve acknowledgement of component risk levels and improve focus of plant resources and costs to the plant components that are most risk significant. Initial plans are in development to implement AI in targeted areas that do not compromise safety and for which the AI tool output can be assessed and managed. The use of AI for smart search and summary processing is already in progress.

REFERENCES

[1] INTERNATIONAL ATOMIC ENERGY AGENCY, Sustaining Operational Excellence at Nuclear Power Plants, IAEA Nuclear Energy Series No. NR-G-3.1, IAEA, Vienna (2022).

[2] INTERNATIONAL ATOMIC ENERGY AGENCY, Institutional Strength in Depth in the Nuclear Industry to Sustain Operational Excellence, IAEA-TECDOC-2038, IAEA, Vienna (2023).

[3] WORLD ASSOCIATION OF NUCLEAR OPERATORS, Principles for Effective Operational Decision Making, WANO GL-2002-01 Rev. 1, Guideline GL-2002-01 Rev. 1, WANO, London (2018).

[4] INTERNATIONAL ATOMIC ENERGY AGENCY, Preparedness and Response for a Nuclear or Radiological Emergency, General Safety Requirements GSR Part 7, IAEA, Vienna (2015).

[5] INTERNATIONAL ATOMIC ENERGY AGENCY, Arrangements for Preparedness for a Nuclear or Radiological Emergency, General Safety Guides GS-G-2.1, INTERNATIONAL ATOMIC ENERGY AGENCY, Vienna (2007).

[6] WIKIPEDIA, DIKW Pyramid, https://en.wikipedia.org/wiki/DIKW_pyramid.

[7] INTERNATIONAL NUCLEAR SAFETY ADVISORY GROUP, Maintaining the Design Integrity of Nuclear Installations throughout Their Operating Life, INSAG Series INSAG-19, IAEA, Vienna (2003).

[8] COMES, T., ADROT, A., RIZZA, C., "Decision Making Under Uncertainty (Section 4.2)", Science for Disaster Risk Management 2017 : Knowing Better and Losing Less, JRC Ispra, Brussels (2017).

[9] INTERNATIONAL ATOMIC ENERGY AGENCY, Considerations on Performing Integrated Risk Informed Decision Making, IAEA-TECDOC-1909, IAEA, Vienna (2020).

[10] OECD NUCLEAR ENERGY AGENCY, Government and Nuclear Energy, Nuclear Development NEA No. 5270, OECD-NEA, Paris (2004).

[11] THE BRIDGESPAN GROUP, Decision-Making Best Practices Checklist, The Bridgespan Group, New York.

[12] INTERNATIONAL ATOMIC ENERGY AGENCY, Management of Nuclear Power Plant Projects, IAEA Nuclear Energy Series No. NG-T-1.6, IAEA, Vienna (2020).

[13] INTERNATIONAL ATOMIC ENERGY AGENCY, Managing Change in Nuclear Utilities, IAEA-TECDOC-1226, IAEA, Vienna (2001).

[14] TVERSKY, A., KAHNEMAN, D., Judgment under Uncertainty: Heuristics and Biases, Science, New Series **185** 4157 (1974) 1124.

[15] SIBONY, OLIVIER, You Are about to Make a Terrible Mistake, (2020).

[16] SRI, V., What's Right for Your Company? Decision Making in 3 Different Organizational Structures.

[17] SHELDON, R., BURNS, E., BRUSH, K., What Is a Decision-Making Process?, *Tech Target Business Analytics*.

[18] INTERNATIONAL ATOMIC ENERGY AGENCY, Nuclear Energy Basic Principles, IAEA Nuclear Energy Series No. NE-BP, IAEA, Vienna (2009).

[19] INTERNATIONAL NUCLEAR SAFETY ADVISORY GROUP, A Framework for an Integrated Risk Informed Decision Making Process, INSAG-25, IAEA, Vienna (2011) 39 pp.

[20] INSTITUTE OF NUCLEAR POWER OPERATORS, Principles for Excellence in Integrated Risk Management, INPO 15-011, INPO, Atlanta (2015).

[21] B. JOHN GARRICK INSTITUTE FOR THE RISK SCIENCES, THE NUCLEAR RISK RESEARCH CENTER, CENTRAL, RESEARCH INSTITUTE OF THE ELECTRIC POWER INDUSTRY, Risk-Informed Decision Making: A Survey of United States Experience, UCLA Engineering and CREPI (2017).

[22] UNITED STATES NUCLEAR REGULATORY COMMISSION, Regulatory Guide 1.174 Rev. 3 An Approach for Using Probabilistic Risk Assessment in Risk-Informed Decisions on Plant-Specific Changes to the Licensing Basis, USNRC, Washington (2018).

[23] INTERNATIONAL ATOMIC ENERGY AGENCY, IAEA Safety Glossary Terminology Used in Nuclear Safety, Nuclear Security, Radiation Protection and Emergency Preparedness and Response 2022 (Interim) Edition, Non-serial Publications IAEA/NSS/GLO, IAEA, Vienna (2022) 246 pp.

[24] INTERNATIONAL ATOMIC ENERGY AGENCY, Leadership and Management for Safety, IAEA Safety Standards Series No. GSR Part 2, IAEA, Vienna (2016).

[25] INTERNATIONAL ATOMIC ENERGY AGENCY, The Management System for Nuclear Installations, IAEA Safety Standards Series No. GS-G-3.5, IAEA, Vienna (2009) 139 pp.

[26] WORLD ASSOCIATION OF NUCLEAR OPERATORS, Traits of a Healthy Nuclear Safety Culture, WANO PL 2013-1, 1st edn, WANO, London (2014).

[27] INTERNATIONAL ATOMIC ENERGY AGENCY, Safety of Nuclear Power Plants: Design, IAEA Safety Standards Series No. SSR-2/1 (Rev. 1), IAEA, Vienna (2016) 71 pp.

[28] WORLD NUCLEAR ASSOCIATION, Implementation of the Design Authority Within a Nuclear Operating Organization, No. 2015/002, London (2015).

[29] INTERNATIONAL ATOMIC ENERGY AGENCY, WORLD ASSOCIATION OF NUCLEAR OPERATORS, Guideline Independent Oversight, GL 2018-01, WANO, London (2018).

[30] NUCLEAR ENERGY INSTITUTE, Operability Determination, NEI 18-03, Nuclear Energy Institute, Washington, D.C. (2019).

[31] HARVARD BUSINESS REVIEW, KAHNEAMAN, D., CHARAN, R., SIBONY, O., HBR's 10 Must Reads on Making Smart Decisions (with Featured Article "Before You Make That Big Decision..." by Daniel Kahneman, Dan Lovallo, and Olivier Sibony), Harvard Business Review Press, Boston (2013).

[32] UNITED STATES NUCLEAR REGULATORY COMMISSION, Cognitive Basis for Human Reliability Analysis (NUREG-2114), USNRC, Washington (2016).

[33] INTERNATIONAL ATOMIC ENERGY AGENCY, Application of the Management System for Facilities and Activities, IAEA Safety Standards Series No. GS-G-3.1, IAEA, Vienna (2006) 105 pp.

[34] INTERNATIONAL ATOMIC ENERGY AGENCY, The Management System for the Processing, Handling and Storage of Radioactive Waste, IAEA Safety Standards Series No. GS-G-3.3, IAEA, Vienna (2008).

[35] INTERNATIONAL ATOMIC ENERGY AGENCY, The Management System for the Disposal of Radioactive Waste, IAEA Safety Standards Series No. GS-G-3.4, IAEA, Vienna (2008).

[36] INTERNATIONAL ATOMIC ENERGY AGENCY, The Management System for the Safe Transport of Radioactive Material, IAEA Safety Standards Series No. TS-G-1.4, IAEA, Vienna (2008).

[37] INTERNATIONAL ATOMIC ENERGY AGENCY, Development and Implementation of a Process Based Management System, IAEA Nuclear Energy Series No. NG-T-1.3, IAEA, Vienna (2015) 57 pp.

[38] INTERNATIONAL ATOMIC ENERGY AGENCY, Guide to Knowledge Management Strategies and Approaches in Nuclear Energy Organizations and Facilities, IAEA Nuclear Energy Series No. NG-G-6.1, IAEA, Vienna (2022).

[39] INTERNATIONAL ATOMIC ENERGY AGENCY, Comparative Analysis of Methods and Tools for Nuclear Knowledge Preservation, IAEA Nuclear Energy Series No. NG-T-6.7, IAEA, Vienna (2011).

[40] INTERNATIONAL ATOMIC ENERGY AGENCY, Knowledge Management and Its Implementation in Nuclear Organizations, IAEA Nuclear Energy Series No. NG-T-6.10, IAEA, Vienna (2016) 52 pp.

[41] INTERNATIONAL ATOMIC ENERGY AGENCY, Development of Knowledge Portals for Nuclear Power Plants, IAEA Nuclear Energy Series No. NG-T-6.2, IAEA, Vienna (2008).

[42] INTERNATIONAL ATOMIC ENERGY AGENCY, Web Harvesting for Nuclear Knowledge Preservation, IAEA Nuclear Energy Series No. NG-T-6.6, IAEA, Vienna (2008).

[43] INTERNATIONAL ATOMIC ENERGY AGENCY, Case Studies on the Development of a Comprehensive Report to Support the Decision Making Process for a Nuclear Power Programme, IAEA-TECDOC-1993, IAEA, Vienna (2022).

[44] INTERNATIONAL ATOMIC ENERGY AGENCY, Knowledge Management for Nuclear Industry Operating Organizations, IAEA-TECDOC-1510, IAEA, Vienna (2006).

[45] INTERNATIONAL ATOMIC ENERGY AGENCY, Knowledge Management Perspectives on Outsourcing in Operating Nuclear Power Plants, IAEA-TECDOC-1884, IAEA, Vienna (2019).

[46] INTERNATIONAL ATOMIC ENERGY AGENCY, Mentoring and Coaching for Knowledge Management in Nuclear Organizations, IAEA-TECDOC-1999, IAEA, Vienna (2022).

[47] INTERNATIONAL ATOMIC ENERGY AGENCY, Planning and Execution of Knowledge Management Assist Visits for Nuclear Organizations, IAEA-TECDOC-1880, IAEA, Vienna (2019).

[48] HERSEY, P. AND BLANCHARD, K.H., Management of Organizational Behavior: Utilizing Human Resources, Prentice Hall, New Jersey, the USA (1969).

[49] INSTITUTE OF NUCLEAR POWER OPERATORS, Leadership and Team Effectiveness Attributes, INPO 15-005, Rev. 1, INPO, Atlanta (2016).

[50] WORLD ASSOCIATION OF NUCLEAR OPERATORS, Nuclear Leadership Effectiveness Attributes, WANO PL 2019-1, WANO, London (2019).

[51] SYED, M., Rebel Ideas: The Power of Diverse Thinking, 2020, John Murray Press.

[52] INTERNATIONAL ATOMIC ENERGY AGENCY, Recruitment, Qualification and Training of Personnel for Nuclear Power Plants, IAEA Safety Standards Series No. NS-G-2.8, IAEA, Vienna (2002).

[53] INTERNATIONAL ATOMIC ENERGY AGENCY, IAEA Tools and Methodologies for Energy System Planning and Nuclear Energy System Assessment, IAEA (2009).

[54] INTERNATIONAL ATOMIC ENERGY AGENCY, IAEA Nuclear Contracting Toolkit, https://nucleus.iaea.org/sites/connect/MSNpublic/Pages/nct/site/index.html.

[55] INTERNATIONAL ATOMIC ENERGY AGENCY, Issues to Improve the Prospects of Financing Nuclear Power Plants, IAEA Nuclear Energy Series No. NG-T-4.1, IAEA, Vienna (2009) 33 pp.

[56] INTERNATIONAL ATOMIC ENERGY AGENCY, Financing of New Nuclear Power Plants, IAEA Nuclear Energy Series No. NG-T-4.2, IAEA, Vienna (2008) 10 pp.

[57] INTERNATIONAL ATOMIC ENERGY AGENCY, Financing Arrangements for Nuclear Power Projects in Developing Countries, Technical Reports Series No. 353, IAEA, Vienna (1993) 189 pp.

[58] INTERNATIONAL ATOMIC ENERGY AGENCY, Managing the Financial Risk Associated with the Financing of New Nuclear Power Plant Projects, IAEA Nuclear Energy Series No. NG-T-4.6, IAEA, Vienna (2017).

[59] WORLD NUCLEAR ASSOCIATION, Nuclear Power Economics and Project Structuring, 2017/001, WNA, London (2017).

[60] INTERNATIONAL ATOMIC ENERGY AGENCY, Procurement Engineering and Supply Chain Guidelines in Support of Operation and Maintenance of Nuclear Facilities, IAEA Nuclear Energy Series No. NP-T-3.21, IAEA, Vienna (2016) 249 pp.

[61] INTERNATIONAL ATOMIC ENERGY AGENCY, Industrial Involvement to Support a National Nuclear Power Programme, Nuclear Energy Series No. NG-T-3.4, IAEA, Vienna (2016) 66 pp.

[62] INTERNATIONAL ATOMIC ENERGY AGENCY, Milestones in the Development of a National Infrastructure for Nuclear Power, Rev. 2, IAEA Nuclear Energy Series No. NG-G-3.1, Rev. 2, IAEA, Vienna (2024).

[63] OECD NUCLEAR ENERGY AGENCY, Third NEA Stakeholder Involvement Workshop on Optimisation in Decision Making: Summary of the Outcomes from the Preparatory Webinars, OECD-NEA, Paris (2023).

[64] SILK, D.M., KATZ, D.A., NILES, S., LIPTON, W., ESG and Sustainability: The Board's Role, *Harvard Law School Forum on Corporate Governance* (2018).

[65] INTERNATIONAL ATOMIC ENERGY AGENCY, Stakeholder Engagement in Nuclear Programmes, IAEA Nuclear Energy Series No. NG-G-5.1, IAEA, Vienna (2021).

[66] INTERNATIONAL ATOMIC ENERGY AGENCY, Communication and Stakeholder Involvement in Environmental Remediation Projects, IAEA Nuclear Energy Series No. NW-T-3.5, IAEA, Vienna (2014) 43 pp.

[67] INTERNATIONAL ATOMIC ENERGY AGENCY, Stakeholder Involvement Throughout the Life Cycle of Nuclear Facilities, IAEA Nuclear Energy Series No. NG-T-1.4, IAEA, Vienna (2011).

[68] INTERNATIONAL NUCLEAR SAFETY ADVISORY GROUP, Stakeholder Involvement in Nuclear Issues, INSAG-20, IAEA, Vienna (2006).

[69] SARIN, R.K., "Multi-attribute Utility Theory", Encyclopedia of Operations Research and Management Science (GASS, S.I., FU, M.C., Eds), (2016).

[70] THOMAS, S., How to make a decision: The analytic hierarchy process, European Journal of Operational Research **48** 1 (1990) 9.

[71] ROGERS, M., BRUEN, M., MAYSTRE, L.-Y., "The Electre Methodology", ELECTRE and Decision Support: Methods and Applications in Engineering and Infrastructure Investment, Springer US, Boston, MA (2000) 45–85.

[72] BRANS, J.P., MARESCHAL, B., VINCKE, Ph., Promethee: A new family of outranking methods in multicriteria analysis., OR'84, (1984) 408.

[73] U.S.NRC, Updated Implementation Guidelines for SSHAC Hazard Studies, NUREG 2213, (2018).

[74] INTERNATIONAL ATOMIC ENERGY AGENCY, Implementation and Effectiveness of Actions Taken at Nuclear Power Plants Following the Fukushima Daiichi Accident, IAEA-TECDOC-1930, IAEA, Vienna (2020).

[75] CHARLES R. SCHWENK, The Use of Devil's Advocates in Strategic Decision-Making, College of Commerce and Business Administration, Bureau of Economic and Business Research, University of Illinois, Urbana-Champaign (1984).

[76] RECOMMENDATIONS OF THE INTERNATIONAL COMMISSION ON RADIOLOGICAL PROTECTION, ICRP Publication 9, Recommendations of the ICRP, Recommendations of the International Commission on Radiological Protection, Pergamon Press, Oxford (1965).

[77] RECOMMENDATIONS OF THE INTERNATIONAL COMMISSION ON RADIOLOGICAL PROTECTION, ICRP Publication 22, Recommendations of the ICRP, Implications of Commission Recommendations That Doses Be Kept as Low as Readily Achievable, Pergamon Press, Oxford (1973).

[78] HER EXCELLENCY THE GOVERNOR GENERAL IN COUNCIL, Directive to the Canadian Nuclear Safety Commission Regarding the Health of Canadians SOR/2007-282, (2007).

[79] UNITED STATES NUCLEAR REGULATORY COMMISSION, Regulatory Analysis Technical Evaluation Handbook (Final Report), NUREG/BR 0184, USNRC, Washington (1997).

[80] UNITED STATES NUCLEAR REGULATORY COMMISSION, Reassessment of NRC's Dollar Per Person-Rem Conversion Factor Policy (Final Report), NUREG 1530, Revision 1, USNRC, Washington.

[81] UNITED STATES NUCLEAR REGULATORY COMMISSION, A Handbook for Value-Impact Assessment, NUREG/CR-3568, USNRC, Washington (1983).

[82] UNITED STATES NUCLEAR REGULATORY COMMISSION, 10 CFR Part 50 Section 109 Backfitting.

[83] WORLD ECONOMIC FORUM, Artificial Intelligence Will Transform Decision-Making. Here's How, Emerging Technologies, https://www.weforum.org/stories/2023/09/how-artificial-intelligence-will-transform-decision-making/.

[84] CANADIAN NUCLEAR SAFETY COMMISSION, UNITED KINGDOM OFFICE OF NUCLEAR REGULATION, UNITED STATES NUCLEAR REGULATORY COMMISSION, Considerations For Developing Artificial Intelligence Systems in Nuclear Applications, CNSC, ONR and US NRC, Ottawa, Liverpool and Rockville, MD (2024).

[85] MATSUZAWA, T., IIO, Y., The Reasons Why We Failed to Anticipate M9 Earthquakes in Northeast Japan, Proceedings of the International Symposium on Engineering Lessons

Learned from the 2011 Great East Japan Earthquake, March 1-4, 2012, Tokyo, Japan, JAEE, Tokyo (2012).

[86] THE INDEPENDENT INVESTIGATION FUKUSHIMA NUCLEAR ACCIDENT, The Fukushima Daiichi Nuclear Power Station Disaster – Investigating the Myth and Reality, BRICKER, M.K., Ed, Routledge, Abingdon, UK and New York (2014).

[87] INSTITUTE OF NUCLEAR POWER OPERATORS, Lessons Learned from the Nuclear Accident at the Fukushima Daiichi Nuclear Power Station, Special Report INPO 11-005, Addendum, INPO, Atlanta (2012).

[88] TOKYO ELECTRIC POWER COMPANY, INC., Fukushima Nuclear Accident Summary & Nuclear Safety Reform Plan, TEPCO, Tokyo, Japan (2013).

[89] INTERNATIONAL ATOMIC ENERGY AGENCY, Development and Application of Level 1 Probabilistic Safety Assessment for Nuclear Power Plants (Rev. 1), IAEA Safety Standards Series No. SSG-3, IAEA, Vienna (2024) 239 pp.

[90] INTERNATIONAL ATOMIC ENERGY AGENCY, Development and Application of Level 2 Probabilistic Safety Assessment for Nuclear Power Plants, IAEA Safety Standards Series No. SSG-4, IAEA, Vienna (2010).

[91] INTERNATIONAL ATOMIC ENERGY AGENCY, Review of Probabilistic Safety Assessments by Regulatory Bodies, IAEA Safety Reports Series No. 25, IAEA, Vienna (2002) 24 pp.

[92] INTERNATIONAL ATOMIC ENERGY AGENCY, A Framework for a Quality Assurance Programme for PSA, IAEA-TECDOC-1101, IAEA, Vienna (1999).

[93] INTERNATIONAL ATOMIC ENERGY AGENCY, Applications of Probabilistic Safety Assessment (PSA) for Nuclear Power Plants., IAEA-TECDOC-1200, IAEA, Vienna (2001).

[94] INTERNATIONAL ATOMIC ENERGY AGENCY, Procedures for Conducting Probabilistic Safety Assessment for Non-Reactor Nuclear Facilities, IAEA-TECDOC-1267, IAEA, Vienna (2002).

[95] INTERNATIONAL ATOMIC ENERGY AGENCY, Determining the Quality of Probabilistic Safety Assessment (PSA) for Applications in Nuclear Power Plants, IAEA-TECDOC-1511, IAEA, Vienna (2006) 178 pp.

[96] INTERNATIONAL ATOMIC ENERGY AGENCY, Level 1 Probabilistic Safety Assessment Practices for Nuclear Power Plants with CANDU Type Reactors, IAEA-TECDOC-1977, IAEA, Vienna (2021) 220 pp.

[97] ASME INTERNATIONAL, ASME/ANS RA-S-1.1–2022 [Revision and Redesignation of ASME/ANS RA-S-2008 (R2019)] Standard for Level 1/Large Early Release Frequency Probabilistic Risk Assessment for Nuclear Power Plant Applications, ASME International, New York (2022).

[98] ASME INTERNATIONAL, RA-S-1.2 Severe Accident Progression and Radiological Release (Level 2) PRA Standard for Nuclear Power Plant Applications for Light Water Reactors (LWRs) -, ASME International, New York (2017).

[99] ASME INTERNATIONAL, RA-S-1.3 - 2017 Standard for Radiological Accident Offsite Consequence Analysis (Level 3 PRA) to Support Nuclear Installation Applications, ASME International, New York (2017).

[100] ASME INTERNATIONAL, RA-S-1.4 - 2021 Probabilistic Risk Assessment Standard for Advanced Non-Light Water Reactor Nuclear Power Plants, ASME International, New York (2021).

[101] CANADIAN STANDARDS ASSOCIATION, Probabilistic Safety Assessment for Nuclear Power Plants, N290.17-17 (R2022), CSA, Toronto (2022).

[102] INTERNATIONAL ELECTROTECHNICAL COMMISSION, IEC 61882:2016 Hazard and Operability Studies (HAZOP Studies) - Application Guide, IEC, Geneva (2016).

[103] AMERICAN PETROLEUM INSTITUTE, API RP 14C Analysis, Design, Installation, and Testing of Safety Systems for Offshore Production Facilities, Eighth Edition, Includes Errata 1 (2018), API (2001).

[104] HEALTH AND SAFETY EXECUTIVE (HSE), Review of Human Reliability Methods, HSE, London (2009).

[105] INTERNATIONAL ATOMIC ENERGY AGENCY, Protection against Internal and External Hazards in the Operation of Nuclear Power Plants, IAEA Safety Standards Series No. SSG-77, IAEA, Vienna (2022) 104 pp.

[106] INTERNATIONAL ATOMIC ENERGY AGENCY, Design of Nuclear Installations Against External Events Excluding Earthquakes, IAEA Safety Standards Series No. SSG-68, IAEA, Vienna (2021) 112 pp.

[107] INTERNATIONAL ATOMIC ENERGY AGENCY, Evaluation of Seismic Safety for Existing Nuclear Installations, IAEA Safety Standards Series No. NS-G-2.13, IAEA, Vienna (2009).

[108] INTERNATIONAL ATOMIC ENERGY AGENCY, Consideration of External Hazards in Probabilistic Safety Assessment for Single Unit and Multi-Unit Nuclear Power Plants, IAEA Safety Reports Series No. 92, IAEA, Vienna (2018).

[109] INTERNATIONAL ATOMIC ENERGY AGENCY, Approaches to Safety Evaluation of New and Existing Research Reactor Facilities in Relation to External Events, IAEA Safety Reports Series No. 94, IAEA, Vienna (2015) 87 pp.

[110] INTERNATIONAL COMMISSION ON RADIOLOGICAL PROTECTION, The 2007 Recommendations of the International Commission on Radiological Protection, Publication 103, Vol. 37, Elsevier (2007).

[111] INTERNATIONAL COMMISSION ON RADIOLOGICAL PROTECTION, The Optimisation of Radiological Protection – Broadening the Process, Publication 101b, Vol. 36, Elsevier (2006).

[112] OECD NUCLEAR ENERGY AGENCY, Adequate Protection after the Fukushima Daiichi Accident; A Constant in a World of Change; Article from the Nuclear Law Bulletin, No. 91, Legal Affairs, OECD-NEA, Paris (2013).

[113] UNITED STATES NUCLEAR REGULATORY COMMISSION, "10 CFR 50.69 - Risk-informed categorization and treatment of structures, systems and components for nuclear power reactors.", Code of Federal Regulations - CFR 10, USNRC, Washington, D.C. (2024).

[114] UNITED STATES NUCLEAR REGULATORY COMMISSION, MODEL APPLICATION FOR PLANT-SPECIFIC ADOPTION OF TSTF-505, REVISION x, "PROVIDE RISK-INFORMED EXTENDED COMPLETION TIMES – RITSTF INITIATIVE 4B, USNC, Washington, D.C. (2011).

[115] NUCLEAR ENERGY INSTITUTE, Risk-Informed Technical Specifications Initiative 4b, Risk-Managed Technical Specifications (RMTS) Guidelines, Industry Guidance Document, NEI 06-09, NEI, Washington, D.C. (2006).

[116]   NUCLEAR ENERGY INSTITUTE, 10 CFR 50.69 SSC Categorization Guideline, NEI 00-04, NEI, Washington, D.C. (2005).

ABBREVIATIONS

AHP	analytical hierarchy process
AI	artificial intelligence
ALARA	as low as reasonably achievable
API	American Petroleum Institute
BN	Bayesian network
CBA	cost-benefit analysis
CDF	core damage frequency
CDT	criteria development team
CRIEPI	Central Research Institute for Electric Power Industry
CSA	Canadian Standards Association
DA	design authority
DG	decision gate
DM	decision making
DMC	decision making centre
DSEP	detailed scoping, estimating, and planning (process)
ECCS	Emergency Core Cooling System
EDG	Emergency Diesel Generator
ELECTRE	ELimination Et Choix Traduisant la réalité
ESG	Environmental, social and governance
EWI	early warning index
EWS	early warning system
FOAK	first-of-a-kind
FMEA	failure mode and effects analysis
GSR	General Safety Requirement
HAZOP	Hazard and Operability Analysis
HRA	human reliability analysis
ICRP	International Commission on Radiological Protection
IEC	Incident and Emergency Centre or International Electrotechnical Commission
INPO	Institute of Nuclear Power Operations

ISID	institutional strength in depth
LERF	large early release frequency
LLM	large language model
LTO	long term operation
MAUT	multi-attribute utility theory
MCDA	multi-criteria decision analysis
MCR	main control room
NPP	nuclear power plant
NRRC	Nuclear Risk Research Centre
ODM	operational decision making
ODMI	operational decision making issues
PSA	probabilistic safety assessment
PSHA	probabilistic seismic hazard analysis
PSR	probabilistic safety review
RICT	risk-informed completion times
RIDM	risk informed decision making
RPV	Reactor Pressure Vessel
SAM	severe accident measures
SSC	structures, systems and components
SSHAC	senior seismic hazard analysis committee
SSR	Specific Safety Requirement
STC	scientific and technical council
TS	technical specification
USNRC	United States Nuclear Regulatory Commission

CONTRIBUTORS TO DRAFTING AND REVIEW

Boring, R.	Idaho National Laboratory, United States of America
Grigoryan, A.	Armenian Nuclear Power plant, Armenia
Harrison, M.	EDF Energy UK Nuclear Generation, United Kingdom
Jager, G.	Consultant, Canada
Kawano, A.	International Atomic Energy Agency
Katsman, A.	Rosenergoatom, Russian Federation
McDermott, B.J.	Tennessee Valley Authority, United States of America
Miao, Y.	China National Nuclear Power Co Ltd, China
Moore, J.H.	Consultant, Canada
Omoto, A.	Consultant, Japan
Russell, P.	INPO, United States of America
Salvatores, S.	EDF, France
Schrader, K.J.	Pressurized Water Reactor Owners Group, United States of America
Schwalbe, D.	Consultant, Germany
Tuomisto, H.	Consultant, Finland

Technical Meeting

Vienna, Austria: 5–8 November 2024

Consultants Meetings

Vienna, Austria: 20–22 June 2023, 14–17 November 2023, 23–26 April 2024, 3–5 September 2024,
5–8 November 2024

CONTACT IAEA PUBLISHING

Feedback on IAEA publications may be given via the on-line form available at:
www.iaea.org/publications/feedback

This form may also be used to report safety issues or environmental queries concerning IAEA publications.

Alternatively, contact IAEA Publishing:

Publishing Section
International Atomic Energy Agency
Vienna International Centre, PO Box 100, 1400 Vienna, Austria
Telephone: +43 1 2600 22529 or 22530
Email: sales.publications@iaea.org
www.iaea.org/publications

Priced and unpriced IAEA publications may be ordered directly from the IAEA.

ORDERING LOCALLY

Priced IAEA publications may be purchased from regional distributors and from major local booksellers.

Printed and bound by CPI Group (UK) Ltd, Croydon, CR0 4YY

06/07/2026

02160600-0006